Concise Vector Analysis

C. J. Eliezer

Dover Publications, Inc.
Mineola, New York

Bibliographical Note

This Dover edition, first published in 2015, is an unabridged republication of the work originally published in 1963 by Pergamon Press Ltd, Oxford.

Library of Congress Cataloging-in-Publication Data

Eliezer, C. J.
 Concise vector analysis / C.J. Eliezer. — Dover edition.
    pages cm
 Originally published: Oxford : Pergamon Press Ltd, 1963.
 Includes index.
 ISBN 978-0-486-80280-0
 ISBN 0-486-80280-9
  1. Vector analysis. I. Title.
QA261.E45 2015
515'.63—dc23

2015026001

Manufactured in the United States by RR Donnelley
80280901   2015
www.doverpublications.com

CONTENTS

Preface   v

Chapter 1. *Vectors and Vector Addition*

1.1 Vectors	1
1.2 Representation of vectors	2
1.3 Vector addition	3
1.4 $-\mathbf{a}, \mathbf{O}, \lambda\mathbf{a}$	5
1.5 Resolution of a vector	7
1.6 Point dividing AB in the ratio $m:n$	10
1.7 Centroid or mean centre of n points	12
Exercises	14

Chapter 2. *Products of Vectors*

2.1 Scalar product	16
2.2 Vector product	22
2.3 Triple products	28
2.4 Mutual moment of two lines	30
2.5 Positive and negative triads	31
Exercises	21, 28 and 32

Chapter 3. *Vector Calculus*

3.1 Vector function of a scalar	38
3.2 Unit tangent vector $\mathbf{T}$	42
3.3 Functions of a vector	44
3.4 Map of a field	45
3.5 Directional derivative	48
3.6 Gradient vector	50
Exercises	55

CONTENTS

Chapter 4. *Vector Calculus (Continued)*

4.1	Line integrals	58
4.2	Line integral of grad ϕ	65
4.3	Surface integrals	71
4.4	Volume integrals	78
4.5	Divergence	81
4.6	Gauss's transformation	85
4.7	Curl **A**	88
4.8	Stokes' theorem	92
4.9	The operator ∇	94
4.10	The Laplacian operator	96
4.11	Orthogonal curvilinear coordinates	96
	Exercises	98

Chapter 5. *Some Applications*

5.1	Equivalence of force systems	103
5.2	Poinsot's Central Axis	107
5.3	Space-curve	113
5.4	Infinitesimal rotations. Angular velocity	117
5.5	Angular velocity of a rigid body	119
5.6	Gauss's theorem	124
5.7	Gravitational potential	130
5.8	Equipotential surfaces	140
5.9	Green's theorems	142
	Exercises	145

Index   151

PREFACE

THIS book aims at presenting concisely an introductory account of the methods and techniques of vector analysis. These methods are now accepted as indispensable tools in mathematics, and also in sciences such as physics or engineering. The first two chapters deal with vector algebra, the next two with vector calculus, and the last with some standard applications. The aim has been to provide a simple presentation, keeping the physical ideas to the forefront, and emphasising ease of understanding rather than mathematical rigour. Each chapter contains illustrative examples, as well as exercises, most of which are taken from old University examination papers. I am grateful to the Universities of Cambridge, Ceylon, London and Oxford for permission to use questions from their examination papers. The following abbreviations have been used:

	N.S.	Natural Science Tripos Part I
	P.N.S.	Preliminary to Natural Sciences
	M.T.	Mathematical Tripos Part I
	P.M.	Preliminary in Mathematics
	M.T.II	Mathematical Tripos Part II
	G.	General
	G.I	General (Part I)
	G.II	General (Part II)
	S.	Special

The book is based on lectures given by the author for many years in the University of Ceylon. The idea of writing a book on this subject originated from the need in the Universities in Ceylon for mathematical books in Sinhala and Tamil. It is a pleasure to make these available in the English language to a wider group of students of mathematics, physics and engineering in universities and technical colleges.

Chapter 1
Vectors and Vector Addition

1.1. Vectors

MATHEMATICS has played an important part in the advance of science, and mathematical language has become essential for the formulation of the laws of science. Numbers, functions, vectors, tensors, spinors, matrices are examples of mathematical entities which occur in various branches of applied mathematics.

The physical quantities with which we are concerned here may be divided into two groups: (*a*) scalars, (*b*) vectors. A scalar requires only its magnitude for its specification. For example, temperature, mass, density, volume and energy are scalars. Vectors require both magnitude and direction for their specification. Force, displacement, velocity, acceleration and momentum are vectors.

Vectors may be classified in different ways. In one elementary classification we have three types of vectors distinguished by their effects:

(i) *Unlocalized or free vector* which has magnitude and direction but no particular position associated with it, e.g. the moment of a couple.

(ii) *Sliding vector* or vector localized along a straight line, e.g. force acting on a rigid body (though we speak of a force as acting at a point, by the principle of transmissibility of force, it is the line of action and not the point of application that is needed to specify the force).

(iii) *Tied vector* or vector localized at a point, e.g. electric field.

In the early part of this book whenever we speak of a vector without stating its classification we mean a free vector. Two vectors of the same magnitude and direction, but acting along parallel lines (or along the same line) will be considered as equal or identical. Thereafter other types of vectors are considered.

Strictly speaking, every quantity which has a magnitude and a direction is not necessarily a vector. A vector is defined as a quantity with a magnitude and a direction, and which obeys the same addition rule as displacements, that is, the rule known as the *parallelogram law* of addition. (As an example of a quantity which has magnitude and direction but is not a vector, consider rotations of a rigid body through finite angles. These have magnitudes and directions, but do not obey the parallelogram law of addition, and are not vectors.)

A vector is denoted by symbols in bold type such as **a, b, P, Q** or by letters with arrows written above such as $\overrightarrow{OA}, \overrightarrow{AB}$.

1.2. Representation of vectors

A vector may be represented by a directed line segment, the direction of the line indicating the direction of the vector and the length of segment representing, in an appropriate scale of units, the magnitude of the vector. Parallel segments of the same length will represent equal or identical vectors. Since the exact position of vectors is not important and parallel segments represent equal vectors, it is convenient to represent vectors in a diagram by line segments starting from the same origin, say *O*. Thus the segment

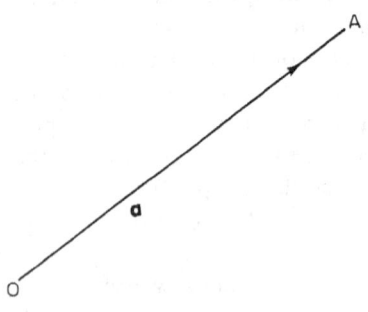

Fig. 1

$\overrightarrow{OA}$ represents a vector **a**, where the direction of $\overrightarrow{OA}$ is the same as the direction of the vector **a**, and the length $\overrightarrow{OA}$ in a suitable scale of units is equal to the magnitude of the vector **a**. We denote the magnitude of the vector **a** by the symbol *a* written in italics or by the symbol |**a**| which is read as "modulus of **a**" or briefly as "mod **a**".

1.3. Vector addition

Let $\overrightarrow{OA}$, $\overrightarrow{OB}$ represent two vectors **a** and **b** respectively. We complete the parallelogram *OACB*, which has *OA*, *OB* as adjacent edges.

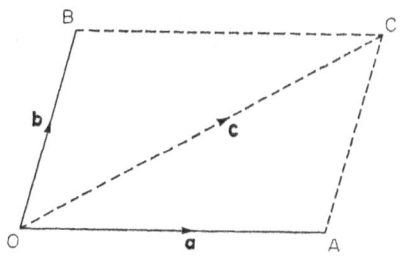

Fig. 2

$\overrightarrow{OA} = \mathbf{a}$, $\overrightarrow{OB} = \mathbf{b}$. Then $\overrightarrow{AC} = \overrightarrow{OB} = \mathbf{b}$, $\overrightarrow{BC} = \overrightarrow{OA} = \mathbf{a}$. Let $\overrightarrow{OC} = \mathbf{c}$. The line segment $\overrightarrow{OA}$ represents the displacement of a particle from *O* to *A*, and $\overrightarrow{AC}$ the displacement from *A* to *C*. The effect of these two displacements in succession is the same as that of a single displacement from *O* to *C*. Hence we write

$$\overrightarrow{OA} + \overrightarrow{AC} = \overrightarrow{OC},$$

that is
$$\mathbf{a} + \mathbf{b} = \mathbf{c}, \tag{1}$$

where + denotes vectorial addition.

This is the triangle rule for vector addition.

We may state the rule in an alternative way, namely that the sum of two vectors **a** and **b** is the vector **c**, which is represented by the diagonal OC of the parallelogram of which OA and OB are adjacent edges. We write

$$\overrightarrow{OB} + \overrightarrow{BC} = \overrightarrow{OC},$$

that is,
$$\mathbf{b} + \mathbf{a} = \mathbf{c},$$

we see that
$$\mathbf{a} + \mathbf{b} = \mathbf{b} + \mathbf{a}. \tag{2}$$

This shows that vector addition is commutative, and the order of addition may be interchanged.

We consider also the associative law of addition. Take three vectors **P**, **Q**, **R** and represent them by $\overrightarrow{AB}$, $\overrightarrow{BC}$, $\overrightarrow{CD}$ respectively. Construct **P** + **Q** and **Q** + **R** and then (**P** + **Q**) + **R** and **P** + (**Q** + **R**).

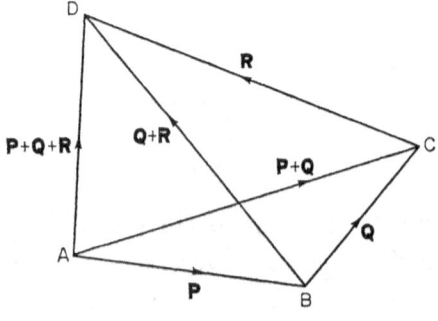

Fig. 3

$$(\mathbf{P} + \mathbf{Q}) + \mathbf{R} = (\overrightarrow{AB} + \overrightarrow{BC}) + \overrightarrow{CD} = \overrightarrow{AC} + \overrightarrow{CD} = \overrightarrow{AD}.$$
$$\mathbf{P} + (\mathbf{Q} + \mathbf{R}) = \overrightarrow{AB} + (\overrightarrow{BC} + \overrightarrow{CD}) = \overrightarrow{AB} + \overrightarrow{BD} = \overrightarrow{AD}.$$

Hence
$$(\mathbf{P} + \mathbf{Q}) + \mathbf{R} = \mathbf{P} + (\mathbf{Q} + \mathbf{R}). \tag{3}$$

VECTORS AND VECTOR ADDITION 5

The brackets which have been used to show the order of the operations may therefore be omitted, and the sum on either side be written as

$$\mathbf{P} + \mathbf{Q} + \mathbf{R}.$$

This result may be extended to define the sum of any number of vectors $\mathbf{P}, \mathbf{Q}, \ldots, \mathbf{W}$.

Suppose we show in a diagram, Fig. 4(a), the vectors $\mathbf{P}, \mathbf{Q}, \ldots, \mathbf{W}$ drawn from an origin O. We draw a polygon $AB \ldots K$ in

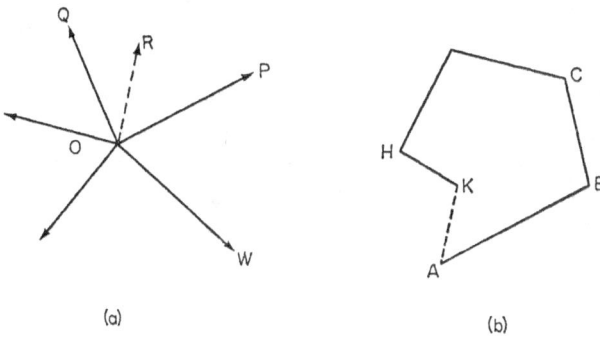

Fig. 4

which the vectors are represented in a suitable scale by directed segments $\overrightarrow{AB}, \overrightarrow{BC} \ldots, \overrightarrow{HK}$, which are placed in succession, with $\overrightarrow{AB} = \mathbf{P}, \overrightarrow{BC} = \mathbf{Q}, \ldots \overrightarrow{HK} = \mathbf{W}$. Join AK. Then $\overrightarrow{AK}$ represents in the same scale the vector sum

$$\mathbf{P} + \mathbf{Q} + \ldots + \mathbf{W}.$$

If the sum is denoted by $\mathbf{R}$, then $\mathbf{R} = \overrightarrow{AK}$. $\mathbf{R}$ is called the *resultant* vector, and $\mathbf{P}, \mathbf{Q}, \ldots$ are called the *components of* $\mathbf{R}$.

1.4. $-\mathbf{a}$, $\mathbf{O}$, $\lambda \mathbf{a}$

The symbol $-\mathbf{a}$ is used to indicate a vector which has the same magnitude as $\mathbf{a}$ but is opposite in direction. If $\mathbf{a} = \overrightarrow{AB}$, then $\overrightarrow{BA} = -\mathbf{a}$

A vector whose terminal points coincide is called a *null* vector and is denoted by the symbol **O**. We see that

$$\mathbf{a} + (-\mathbf{a}) = \overrightarrow{AB} + \overrightarrow{BA} = \overrightarrow{AA} = \mathbf{O}. \tag{4}$$

Also

$$\mathbf{a} + \mathbf{O} = \overrightarrow{AB} + \overrightarrow{BB} = \overrightarrow{AB} = \mathbf{a}. \tag{5}$$

The sum of the vectors **a** and $-\mathbf{b}$, that is $\mathbf{a} + (-\mathbf{b})$, may be written $\mathbf{a} - \mathbf{b}$. In Fig. 2, $\overrightarrow{BO} = -\mathbf{b}$, and $\mathbf{a} + (-\mathbf{b}) = \overrightarrow{BC} + \overrightarrow{BO} = \overrightarrow{BA}$, by the parallelogram law. Hence $\overrightarrow{BA}$ represents $\mathbf{a} - \mathbf{b}$.

If λ is a positive scalar, and **a** a vector then $\lambda\mathbf{a}$ is defined to be the vector which has the same direction as **a** but whose magnitude is λ times that of **a**. If $\overrightarrow{OA}$ represents **a**, and $\overrightarrow{OA'}$ represents $\lambda\mathbf{a}$, then the points O, A, A' lie on a line and their distances are such that $OA'/OA = \lambda$.

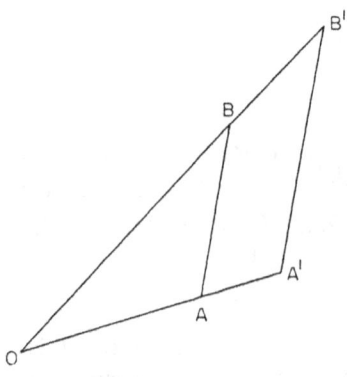

Fig. 5

This multiplication of a vector by a scalar obeys the distributive law.

That is,

$$\lambda(\mathbf{a} + \mathbf{b}) = \lambda\mathbf{a} + \lambda\mathbf{b}. \tag{6}$$

This may be verified by using properties of similar triangles. Let $\vec{OA} = \mathbf{a}$, $\vec{AB} = \mathbf{b}$, then $\vec{OB} = \mathbf{a} + \mathbf{b}$. Let A' and B' be such that

$$OA'/OA = OB'/OB = \lambda.$$

Then $A'B'$ is parallel to AB, and the two triangles OAB, $OA'B'$ are similar.

$$\vec{OA'} = \lambda\mathbf{a}, \ \vec{A'B'} = \lambda\mathbf{b}, \ \vec{OB'} = \lambda\vec{OB} = \lambda(\mathbf{a} + \mathbf{b}).$$

Also

$$\vec{OB'} = \vec{OA'} + \vec{A'B'} = \lambda\mathbf{a} + \lambda\mathbf{b}.$$

Hence we obtain eqn. (6).

We see that the rules for addition and subtraction of vectors, and multiplication by a scalar of vectors agree with the laws of ordinary algebra.

1.5. Resolution of a vector

Let Ox, Oy, Oz be three lines mutually at right angles. These lines may be used as the three axes of a system of Cartesian rectangular coordinates or as a frame of reference.

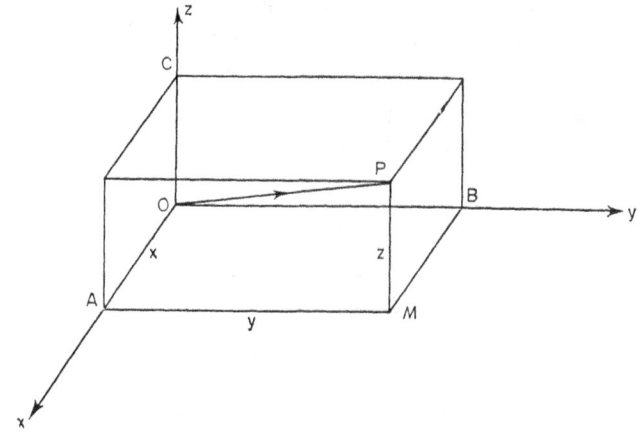

Fig. 6

From a given point P, draw PM normal to the xy-plane (that is, the plane containing the axes Ox, Oy). Draw MA perpendicular to Ox. Let $OA = x$, $AM = y$, $MP = z$. Complete the rectangular solid with sides of length x, y and z shown in the diagram. OA, OB, OC, that is, x, y and z, are the projections of OP on the three axes, respectively. x, y, z are said to be the coordinates of the point P.

Vectorially,

$$\overrightarrow{OP} = \overrightarrow{OA} + \overrightarrow{AM} + \overrightarrow{MP}$$
$$= \overrightarrow{OA} + \overrightarrow{OB} + \overrightarrow{OC}.$$

$\overrightarrow{OA}$, $\overrightarrow{OB}$, $\overrightarrow{OC}$ are the components of $\overrightarrow{OP}$.

When a vector $\overrightarrow{OP}$ is resolved into three mutually perpendicular components, such as $\overrightarrow{OA}$, $\overrightarrow{OB}$, $\overrightarrow{OC}$ (or into two perpendicular components $\overrightarrow{OA}$, $\overrightarrow{OB}$ when OP is in the plane AOB), then each component is called the *resolved part* or *resolute* of the vector in the corresponding direction.

It is convenient to have a notation for the three unit vectors (that is vectors of unit magnitude) along the three axes. Let **i**, **j**, **k** denote unit vectors along the x-, y- and z- axes respectively. Then $\overrightarrow{OA} = x\mathbf{i}$, $\overrightarrow{OB} = y\mathbf{j}$, and $\overrightarrow{OC} = z\mathbf{k}$, are the resolutes of OP along the three axes. $\overrightarrow{OP}$ is called the *position vector* of the point P relative to the origin O, or simply, the position vector of P. If we denote $\overrightarrow{OP}$ by **r**,

$$\mathbf{r} = x\mathbf{i} + y\mathbf{j} + z\mathbf{k}. \tag{7}$$

Also $OP^2 = OA^2 + AP^2$ (OA, AP being at right angles) $= OA^2 + AM^2 + MP^2$. Hence

$$|\mathbf{r}^2| = x^2 + y^2 + z^2. \tag{8}$$

If OP makes angles α, β, γ with the three axes respectively, $\cos \alpha$, $\cos \beta$, $\cos \gamma$ are called the *direction cosines* of OP. We see that $x = r \cos \alpha$, $y = r \cos \beta$, $z = r \cos \gamma$, and the direction cosines of OP are x/r, y/r, z/r.

Suppose the projections of a vector **a** on the 3 axes are a_x, a_y, a_z. a_x, a_y, a_z are scalars. The projections may be represented in magnitude and direction by $a_x\mathbf{i}$, $a_y\mathbf{j}$, $a_z\mathbf{k}$. Then

$$\mathbf{a} = a_x\mathbf{i} + a_y\mathbf{j} + a_z\mathbf{k}. \tag{9}$$

The magnitudes of the resolutes along the three axes are

$$a_x = |\mathbf{a}|\cos(\mathbf{a}, x), \qquad a_y = |\mathbf{a}|\cos(\mathbf{a}, y), \qquad a_z = |\mathbf{a}|\cos(\mathbf{a}, z) \tag{10}$$

where $(\mathbf{a}, x)$ denotes the angle between the vector **a**, and the x-axis. Also, the magnitude of the vector is

$$|\mathbf{a}| = \sqrt{(a_x^2 + a_y^2 + a_z^2)} \tag{11}$$

and the direction cosines of **a** are

$$\left(\frac{a_x}{|\mathbf{a}|}, \frac{a_y}{|\mathbf{a}|}, \frac{a_z}{|\mathbf{a}|}\right). \tag{12}$$

If a vector **a** has resolutes (a_x, a_y, a_z) and **b** has resolutes (b_x, b_y, b_z), then $\mathbf{a} = a_x\mathbf{i} + a_y\mathbf{j} + a_z\mathbf{k}$, $\mathbf{b} = b_x\mathbf{i} + b_y\mathbf{j} + b_z\mathbf{k}$. Using the rules of eqn. (3) and (6), we obtain

$$\mathbf{a} + \mathbf{b} = (a_x + b_x)\mathbf{i} + (a_y + b_y)\mathbf{j} + (a_z + b_z)\mathbf{k} \tag{13}$$

showing that vectors may be compounded by the rule that the resolved part of $(\mathbf{a} + \mathbf{b})$ along any direction is the sum of the resolved parts along the same direction of **a** and **b**.

Example. Points A and B have position vectors

$$\mathbf{a} = a_1\mathbf{i} + a_2\mathbf{j} + a_3\mathbf{k}, \qquad \mathbf{b} = b_1\mathbf{i} + b_2\mathbf{j} + b_3\mathbf{k}.$$

Find the vector $\overrightarrow{AB}$, the length of AB and the direction cosines of AB.

If O is the origin,

$$\overrightarrow{OA} = \mathbf{a}, \qquad \overrightarrow{OB} = \mathbf{b},$$

and
$$\overrightarrow{AB} = \overrightarrow{AO} + \overrightarrow{OB} = (-\mathbf{a}) + \mathbf{b} = \mathbf{b} - \mathbf{a}.$$

r = length of AB
$= |\mathbf{b} - \mathbf{a}| = \sqrt{[(b_1 - a_1)^2 + (b_2 - a_2)^2 + (b_3 - a_3)^2]}.$

Direction cosines of AB are

$$\frac{b_1 - a_1}{r}, \frac{b_2 - a_2}{r}, \frac{b_3 - a_3}{r}.$$

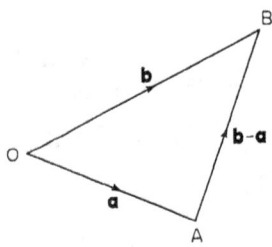

Fig. 7

1.6. Point dividing AB in the ratio $m:n$

If $\mathbf{a}$, $\mathbf{b}$ are the position vectors of two points A and B, the point R which divides AB in the ratio $m:n$ has position vector

$$\mathbf{r} = \frac{n\mathbf{a} + m\mathbf{b}}{n + m}. \tag{14}$$

Proof:
$\overrightarrow{AR} = \mathbf{r} - \mathbf{a}$, $\overrightarrow{RB} = \mathbf{b} - \mathbf{r}$. Since $AR/RB = m/n$, and AR, RB are in the same direction, $n\overrightarrow{AR} = m\overrightarrow{RB}$. Hence

$$n(\mathbf{r} - \mathbf{a}) = m(\mathbf{b} - \mathbf{r})$$

$$\mathbf{r} = \frac{n\mathbf{a} + m\mathbf{b}}{m + n}.$$

VECTORS AND VECTOR ADDITION 11

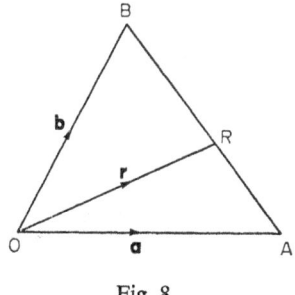

Fig. 8

As a particular case, we obtain that the position vector of the mid-point of AB is

$$\tfrac{1}{2}(\mathbf{a} + \mathbf{b}). \tag{15}$$

Example. If **a, b, c** *are the position vectors of the vertices A, B, C of a triangle, show that the centroid of the triangle has position vector*

$$\tfrac{1}{3}(\mathbf{a} + \mathbf{b} + \mathbf{c}). \tag{16}$$

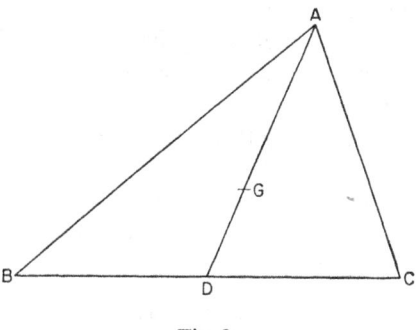

Fig. 9

If D is the mid-point of BC, the centroid G divides AD in the ratio $2:1$.

The position vector of D is $\tfrac{1}{2}(\mathbf{b} + \mathbf{c})$.

Using formula (14), G has position vector

$$\frac{1.\mathbf{a} + 2.(\mathbf{b} + \mathbf{c})/2}{1 + 2} = \frac{\mathbf{a} + \mathbf{b} + \mathbf{c}}{3}.$$

1.7. Centroid or mean centre of n points

Given n points $A_1, A_2, \ldots, A_n$ with position vectors $\mathbf{r}_1, \mathbf{r}_2, \ldots, \mathbf{r}_n$, their *centroid* or *centre of mean position* is defined as the point G with position vector

$$\bar{\mathbf{r}} = \frac{1}{n}(\mathbf{r}_1 + \mathbf{r}_2 + \ldots + \mathbf{r}_n). \tag{17}$$

If $m_1, m_2, \ldots, m_n$ are n real numbers, the point G whose position vector is

$$\bar{\mathbf{r}} = \frac{m_1\mathbf{r}_1 + m_2\mathbf{r}_2 + \ldots + m_n\mathbf{r}_n}{m_1 + m_2 + \ldots + m_n} = \frac{\sum m\mathbf{r}}{\sum m} \tag{18}$$

is defined as the centroid of the given points $A_1, A_2, \ldots, A_n$ with associated numbers $m_1, m_2, \ldots, m_n$ respectively. The point is also called the *weighted mean centre*, the numbers $m_1, m_2, \ldots, m_n$ being the associated weights.

For two points A and B, the point dividing AB in the ratio $m:n$ is the centroid with the associated weights n at A and m at B. (Compare eqn. (14)).

We can show that the centroid is independent of the particular origin O used in defining the vectors. Let O' be a point whose position vector, relative to O is $\mathbf{a}$. Relative to O' as origin, the position vectors of $A_1, A_2, \ldots, A_n$ are $\mathbf{r}_1 - \mathbf{a}, \mathbf{r}_2 - \mathbf{a}, \ldots, \mathbf{r}_n - \mathbf{a}$, and the centroid G is then defined by

$$\overrightarrow{O'G'} = \frac{m_1(\mathbf{r}_1 - \mathbf{a}) + m_2(\mathbf{r}_2 - \mathbf{a}) + \ldots + m_n(\mathbf{r}_n - \mathbf{a})}{m_1 + m_2 + \ldots m_n}$$

$$= \frac{m_1\mathbf{r}_1 + m_2\mathbf{r}_2 + \ldots + m_n\mathbf{r}_n}{m_1 + m_2 + \ldots + m_n} - \mathbf{a}$$

$$= \overrightarrow{OG} - \mathbf{a} = \overrightarrow{O'G}.$$

VECTORS AND VECTOR ADDITION

Hence G and G' coincide, showing that the centroid is independent of the choice of origin.

The centroid of a curve, area or volume may be defined by dividing into a large number of small elements, finding the centroid with associated numbers proportional to the length, area or volume respectively, of each element, and finding the limiting position as n increases indefinitely and each element tends to a point.

Thus

$$\bar{\mathbf{r}} = \frac{\lim \sum \mathbf{r}\delta s}{\lim \sum \delta s} = \frac{\int \mathbf{r}\,\mathrm{d}s}{\int \mathrm{d}s} \tag{19}$$

for a curve,

$$\frac{\int \mathbf{r}\,\mathrm{d}A}{\int \mathrm{d}A} \text{ for an area and } \frac{\int \mathbf{r}\,\mathrm{d}v}{\int \mathrm{d}v} \text{ for a volume.}$$

Example 1. *Show that the centroid G defined by eqn.* (18) *satisfies the relation*

$$m_1\overrightarrow{A_1G} + m_2\overrightarrow{A_2G} + \ldots + m_n\overrightarrow{A_nG} = \mathbf{O}. \tag{20}$$

If G has position vector $\bar{\mathbf{r}}$, $\overrightarrow{A_1G} = \bar{\mathbf{r}} - \mathbf{r}_1$, $\overrightarrow{A_2G} = \bar{\mathbf{r}} - \mathbf{r}_2, \ldots$
L.H.S. of (20) is

$$m_1(\bar{\mathbf{r}} - \mathbf{r}_1) + m_2(\bar{\mathbf{r}} - \mathbf{r}_2) + \ldots + m_n(\bar{\mathbf{r}} - \mathbf{r}_n)$$
$$= (m_1 + m_2 + \ldots + m_n)\bar{\mathbf{r}} - (m_1\mathbf{r}_1 + m_2\mathbf{r}_2 + \ldots + m_n\mathbf{r}_n)$$
$$= \mathbf{O}, \text{ using eqn. (18)}.$$

Example 2. *Given points* $A_1, A_2, \ldots, A_n$ *and associated numbers* $m_1, m_2, \ldots, m_n$ *a point G is defined by the following process: Divide* A_1A_2 *at* G_1 *so that* $m_1A_1G_1 = m_2G_1A_2$. *Then divide* G_1A_3 *at* G_2 *so that* $(m_1 + m_2)G_1G_2 = m_3G_2A_3$. *Continue the process using points* $A_4, A_5, \ldots, A_n$ *successively, and obtaining points* $G_3, G_4, \ldots, G_{n-1}$. *If the last point obtained, namely* G_{n-1}, *is called G, show that G satisfies eqn.* (18).

Since G_1 divides A_1A_2 in the ratio $m_2:m_1$, its position vector is

$(m_1\mathbf{r}_1 + m_2\mathbf{r}_2)/(m_1 + m_2)$. Since G_2 divides G_1A_3 in the ratio $m_3:(m_1 + m_2)$, its position vector is

$$\frac{(m_1 + m_2)\dfrac{(m_1\mathbf{r}_1 + m_2\mathbf{r}_2)}{m_1 + m_2} + m_3\mathbf{r}_3}{(m_1 + m_2) + m_3} = \frac{m_1\mathbf{r}_1 + m_2\mathbf{r}_2 + m_3\mathbf{r}_3}{m_1 + m_2 + m_3}.$$

Proceeding in this way we establish (18).

The symmetry in the formula (18) shows that the point G is independent of the order in which the different points are taken to form the sequence of points $G_1, G_2, \ldots, G_{n-1}$.

Example 3. Given points $A_1, A_2, \ldots, A_n$ and associated numbers $m_1, m_2, \ldots, m_n$, show that a point G defined by the relation

$$m_1\overrightarrow{A_1G} + m_2\overrightarrow{A_2G} + \ldots + m_n\overrightarrow{A_nG} = \mathbf{O} \tag{20}$$

is unique, provided that $m_1 + m_2 + \ldots + m_n \neq 0$.

Suppose that there exists another point G' satisfying this relationship. Then

$$m_1\overrightarrow{A_1G'} + m_2\overrightarrow{A_2G'} + \ldots + m_n\overrightarrow{A_nG'} = \mathbf{O}. \tag{21}$$

Subtracting (20) from (21), we obtain

$$(m_1 + m_2 + \ldots + m_n)\overrightarrow{GG'} = \mathbf{O},$$

showing that if $m_1 + m_2 + \ldots + m_n \neq 0$, G and G' coincide. Hence G is unique.

The case $m_1 + m_2 + \ldots + m_n = 0$ is exceptional, as there is then no centroid at a finite distance.

The centroid G of a system of points with associated weights may thus be defined by eqn. (18) or (20) or by the construction in example 2.

Exercise I

1. If G is the centroid of a triangle ABC and O any point, show that
$$\overrightarrow{GA} + \overrightarrow{GB} + \overrightarrow{GC} = \mathbf{O}$$
$$\overrightarrow{OA} + \overrightarrow{OB} + \overrightarrow{OC} = 3\overrightarrow{OG}.$$

2. If λ, μ are scalar quantities, show that
$$\lambda \overrightarrow{OA} + \mu \overrightarrow{OB} = (\lambda + \mu)\overrightarrow{OC}$$
where C divides AB in the ratio $\mu:\lambda$.

3. A_1, A_2, ... , A_n are n points dividing the circumference of a circle of centre C into n equal parts. If O is any point, simplify
$$\overrightarrow{OA_1} + \overrightarrow{OA_2} + \ldots + \overrightarrow{OA_n}.$$

4. Find the sum of the three vectors which are represented by the diagonals passing through the same vertex of three adjacent faces of a cube.

5. If O is the circumcentre and H the orthocentre of a triangle ABC, show that
$$\overrightarrow{OA} + \overrightarrow{OB} + \overrightarrow{OC} = \overrightarrow{OH},$$
$$\overrightarrow{HA} + \overrightarrow{HB} + \overrightarrow{HC} = 2\overrightarrow{HO},$$
$$\overrightarrow{HA}\tan A + \overrightarrow{HB}\tan B + \overrightarrow{HC}\tan C = \mathbf{O}.$$

6. Find the centroid of
   (a) n points $\mathbf{i}$, $2\mathbf{i}$, $3\mathbf{i}$, ... , $n\mathbf{i}$,
   (b) $2n$ points $\mathbf{i}$, $2\mathbf{i}$, ... , $n\mathbf{i}$, $\mathbf{j}$, $2\mathbf{j}$, ... , $n\mathbf{j}$.

7. Show that the sum of n vectors
$$m_1\overrightarrow{OA_1}, m_2\overrightarrow{OA_2}, \ldots , m_n\overrightarrow{OA_n}$$
is
$$(m_1 + m_2 + \ldots + m_n)\overrightarrow{OG}$$
where G is the centroid of points A_1, A_2, ... , A_n with associated weights m_1, m_2, ... , m_n (Leibnitz's theorem).

8. Show by using vectors that the three straight lines joining the midpoints of opposite sides of a tetrahedron all meet and bisect one another.

9. The position vectors of the three points A, B, C are $\mathbf{a}$, $\mathbf{b}$, $\mathbf{c}$. Find the condition that the point D whose position vector is $\mathbf{d} = \lambda\mathbf{a} + \mu\mathbf{b} + v\mathbf{c}$ is coplanar with A, B, C.

If AB and CD meet in E, and AC and BD in F, find the position vectors of E and F.

Show that the point of intersection of AD and EF has the position vector
$$\frac{1}{1 + \lambda}(\lambda\mathbf{a} + \mathbf{d}). \qquad \text{[Camb. N.S. 1954]}$$

10. $\mathbf{a}$, $\mathbf{b}$, $\mathbf{c}$ are the vectors $\overrightarrow{OA}$, $\overrightarrow{OB}$, $\overrightarrow{OC}$. If there exist numbers α, β, γ not all zero, such that $\alpha + \beta + \gamma = 0$ and $\alpha\mathbf{a} + \beta\mathbf{b} + \gamma\mathbf{c} = \mathbf{O}$, show that A, B and C are collinear. State and prove the converse result. [Camb. P.N.S. 1954]

Chapter 2

Products of Vectors

2.1. Scalar product

GIVEN two numbers a, b we can form their product ab. But given two vectors **a**, **b**, no obvious meaning can be assigned to an expression such as **ab**. We can, however, define certain products of vectors with some special meanings, and these products have various applications which add to the value of the concept of vectors.

Given two vectors **a** and **b** suppose θ is the angle between the directions of these two vectors taken in such a way that $0 \leqslant \theta \leqslant \pi$. The *scalar product* (also called *inner product*) of the vectors **a** and **b** is defined to be the quantity

$$|\mathbf{a}||\mathbf{b}| \cos \theta. \tag{1}$$

This quantity is denoted by **a.b** and is read as "a dot b".

The scalar product is a scalar quantity which has no direction. It is positive when θ is acute, negative when θ is obtuse, and vanishes when $\theta = \pi/2$. Conversely if

$$\mathbf{a.b} = 0 \tag{2}$$

we may infer that either $\mathbf{a} = \mathbf{O}$ or $\mathbf{b} = \mathbf{O}$ or **a** and **b** are perpendicular.

From the definition of **a.b** and the commutative law of ordinary multiplication we see that

$$\mathbf{b.a} = \mathbf{a.b}, \tag{3}$$

that is, scalar product is commutative.

PRODUCTS OF VECTORS 17

The component of **b** in the direction of **a** is denoted by comp$_a$ **b** and has the value $|{\bf b}|\cos\theta$. It is equal to the length of ON, which is the projection of OB on OA, this length being reckoned positive when $\overrightarrow{ON}$ is in the direction of **a**, and negative when $\overrightarrow{ON}$ is opposite to **a**. This may be expressed in terms of scalar product as comp$_a$ **b** = **a**.**b**/$|{\bf a}|$. We note that we may also write

$$\mathbf{a}.\mathbf{b} = |\mathbf{a}|\,\text{comp}_a\,\mathbf{b} = |\mathbf{b}|\,\text{comp}_b\,\mathbf{a}. \qquad (4)$$

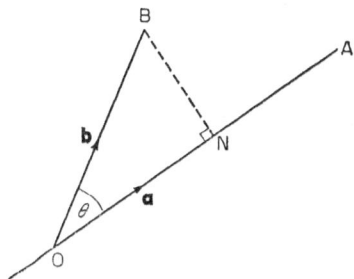

Fig. 1

The component of a vector **a** in a direction whose unit vector is **u** is **a**.**u**.

We may use the symbol $\mathbf{a}^2$ to denote the symbol **a**.**a**. We see that

$$\mathbf{a}^2 = |\mathbf{a}|^2. \qquad (5)$$

If **i**, **j**, **k** are unit vectors along the three axes of coordinates,

$$\left.\begin{array}{lll} \mathbf{i}.\mathbf{i}=1, & \mathbf{i}.\mathbf{j}=0, & \mathbf{i}.\mathbf{k}=0 \\ \mathbf{j}.\mathbf{i}=0, & \mathbf{j}.\mathbf{j}=1, & \mathbf{j}.\mathbf{k}=0 \\ \mathbf{k}.\mathbf{i}=0, & \mathbf{k}.\mathbf{j}=0, & \mathbf{k}.\mathbf{k}=1 \end{array}\right\} \qquad (6)$$

If λ is a scalar

$$(\lambda\mathbf{a}).\mathbf{b} = \lambda\mathbf{a}.\mathbf{b} = \mathbf{a}.(\lambda\mathbf{b}). \qquad (7)$$

The distributive law of multiplication holds for scalar products, that is,

$$\mathbf{a}.(\mathbf{b} + \mathbf{c}) = \mathbf{a}.\mathbf{b} + \mathbf{a}.\mathbf{c}. \tag{8}$$

In Fig. 2, let $\overrightarrow{OA} = \mathbf{a}$, $\overrightarrow{OB} = \mathbf{b}$, $\overrightarrow{BC} = \mathbf{c}$, and let BM, CN be the perpendiculars from B and C on OA.

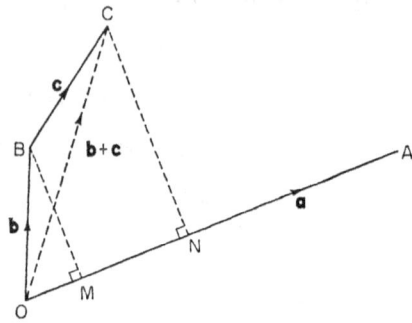

Fig. 2

We see that $\overline{OC} = \mathbf{b} + \mathbf{c}$.
L.H.S. of (8) $= OA.\,\text{comp}_\mathbf{a}\, OC = OA.ON$
R.H.S. of (8) $= OA.OM + OA.MN$
$= OA(OM + MN) = OA.ON$.
This verifies eqn. (8).

These properties of scalar products enable manipulations with scalar products to be carried out by the rules of ordinary algebra. Thus

$$(\mathbf{a} + \mathbf{b}).(\mathbf{c} + \mathbf{d}) = \mathbf{a}.\mathbf{c} + \mathbf{a}.\mathbf{d} + \mathbf{b}.\mathbf{c} + \mathbf{b}.\mathbf{d}. \tag{9}$$

If $\mathbf{a}$ and $\mathbf{b}$ are expressed in terms of their Cartesian components, their scalar product may be obtained in terms of these components in the following way:

PRODUCTS OF VECTORS

$$\mathbf{a}.\mathbf{b} = (a_x\mathbf{i} + a_y\mathbf{j} + a_z\mathbf{k}).(b_x\mathbf{i} + b_y\mathbf{j} + b_z\mathbf{k})$$
$$= a_xb_x\mathbf{i}.\mathbf{i} + a_xb_y\mathbf{i}.\mathbf{j} + a_xb_z\mathbf{i}.\mathbf{k} + a_yb_x\mathbf{j}.\mathbf{i} + a_yb_y\mathbf{j}.\mathbf{j} + a_yb_z\mathbf{j}.\mathbf{k}$$
$$+ a_zb_x\mathbf{k}.\mathbf{i} + a_zb_y\mathbf{k}.\mathbf{j} + a_zb_z\mathbf{k}.\mathbf{k}$$
$$= a_xb_x + a_yb_y + a_zb_z. \tag{10}$$

Also
$$\mathbf{a}^2 = \mathbf{a}.\mathbf{a} = a_x^2 + a_y^2 + a_z^2. \tag{11}$$

Illustration. If the point of application of a force **F**, is given a displacement **s** (without change in magnitude or direction of the force) the work W that is done by the force is, by definition, equal to the product of the magnitude of the force and the component of the displacement in the direction of the force. This work may be expressed conveniently in terms of scalar product as follows:

$$W = |\mathbf{F}| \operatorname{comp}_\mathbf{F} \mathbf{s} = |\mathbf{F}||\mathbf{s}| \cos \theta$$
$$= \mathbf{F}.\mathbf{s}. \tag{12}$$

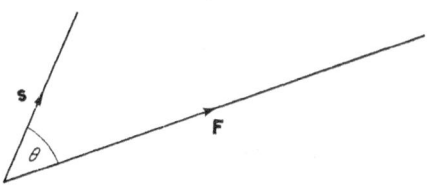

Fig. 3

Example 1. Show that the components of a vector **a** in the directions of the three coordinate axes are $\mathbf{a}.\mathbf{i}, \mathbf{a}.\mathbf{j}, \mathbf{a}.\mathbf{k}$ respectively.

If the components are a_x, a_y, a_z, then $\mathbf{a} = a_x\mathbf{i} + a_y\mathbf{j} + a_z\mathbf{k}$. Taking scalar product of either side with **i** and using $\mathbf{i}.\mathbf{i} = 1$, $\mathbf{i}.\mathbf{j} = 0, \mathbf{i}.\mathbf{k} = 0$, we obtain $\mathbf{a}.\mathbf{i} = a_x$.

Similarly $a_y = \mathbf{a}.\mathbf{j}, a_z = \mathbf{a}.\mathbf{k}$.

Example 2. Using vectors show that the altitudes of a triangle meet in a point.

Let **a**, **b**, **c** be the position vectors of the vertices A, B, C of the triangle, and let **r** be the position vector of the point of intersection H of the altitudes from A and B.

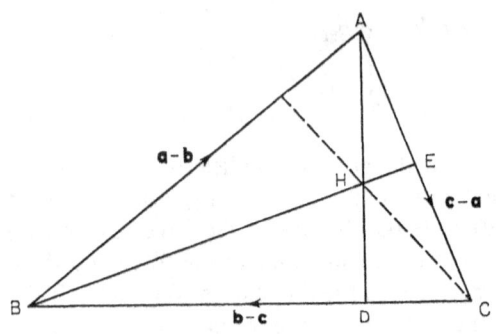

Fig. 4

$$\overrightarrow{AH} = \mathbf{r} - \mathbf{a}, \qquad \overrightarrow{BH} = \mathbf{r} - \mathbf{b}, \qquad \overrightarrow{CH} = \mathbf{r} - \mathbf{c},$$

$$\overrightarrow{BA} = \mathbf{a} - \mathbf{b}, \qquad \overrightarrow{CB} = \mathbf{b} - \mathbf{c}, \qquad \overrightarrow{AC} = \mathbf{c} - \mathbf{a}.$$

Since AH is perpendicular to BC,

$$(\mathbf{r} - \mathbf{a}).(\mathbf{b} - \mathbf{c}) = 0,$$

and since $BH \perp CA$.

$$(\mathbf{r} - \mathbf{b}).(\mathbf{c} - \mathbf{a}) = 0.$$

Adding these two relations, we obtain

$$\mathbf{r}.\mathbf{b} - \mathbf{r}.\mathbf{c} - \mathbf{a}.\mathbf{b} + \mathbf{a}.\mathbf{c} + \mathbf{r}.\mathbf{c} - \mathbf{r}.\mathbf{a} - \mathbf{b}.\mathbf{c} + \mathbf{b}.\mathbf{a} = 0$$

which is seen to be

$$(\mathbf{r} - \mathbf{c}).(\mathbf{b} - \mathbf{a}) = 0,$$

which is the condition that $CH \perp AB$.

Example 3. Solve for **x** the vector equation
$$\mathbf{x} + (\mathbf{x}.\mathbf{b})\mathbf{a} = \mathbf{c} \tag{13}$$
where **a**, **b**, **c** are given vectors such that $1 + \mathbf{a}.\mathbf{b} \neq 0$.

Taking scalar product of (13) with **b** we obtain
$$\mathbf{x}.\mathbf{b} + (\mathbf{x}.\mathbf{b})(\mathbf{a}.\mathbf{b}) = \mathbf{c}.\mathbf{b}$$

Hence
$$\mathbf{x}.\mathbf{b} = \frac{\mathbf{c}.\mathbf{b}}{1 + \mathbf{a}.\mathbf{b}}, \tag{14}$$

using the condition $1 + \mathbf{a}.\mathbf{b} \neq 0$. By substituting for $\mathbf{x}.\mathbf{b}$ in (13) we obtain
$$\mathbf{x} = \mathbf{c} - \frac{\mathbf{c}.\mathbf{b}}{1 + \mathbf{a}.\mathbf{b}}\mathbf{a}. \tag{15}$$

We may verify by direct substitution that the solution (15) does satisfy eqn. (13).

Exercise IIa

1. Show that
$$(\mathbf{a} + \mathbf{b})^2 = \mathbf{a}^2 + 2\mathbf{a}.\mathbf{b} + \mathbf{b}^2$$
$$(\mathbf{a} + \mathbf{b}).(\mathbf{a} - \mathbf{b}) = \mathbf{a}^2 - \mathbf{b}^2.$$

2. Given the vectors
$$\mathbf{a} = -\mathbf{i} + 3\mathbf{j}, \ \mathbf{b} = \mathbf{i} - \mathbf{j} + \mathbf{k}, \ \mathbf{c} = 2\mathbf{i} + \mathbf{j} + \mathbf{k},$$
evaluate

(i) $\mathbf{a}.\mathbf{a}$ (ii) $\mathbf{a}.\mathbf{i}, \mathbf{a}.\mathbf{j}, \mathbf{a}.\mathbf{k}$
(iii) $\text{comp}_\mathbf{a}(\mathbf{b} + \mathbf{c})$ (iv) $\mathbf{a}.\mathbf{b}$ (v) $\cos \widehat{\mathbf{b}\,\mathbf{c}}$

3. Given points $A = (1, 2, -1)$, $B = (0, 0, 1)$, $C = (2, 1, 2)$ find
 (i) the Cartesian components of $\overrightarrow{AB}$
 (ii) the length of AB
 (iii) $\overrightarrow{AB}.\overrightarrow{AC}$
 (iv) the magnitude of the angle CAB.

4. The point of application of the force $\mathbf{F} = (15, 10, 5)$ lb is displaced from a point $A = (3, 0, 2)$ to the point $B = (-6, -1, 3)$. The coordinates being in ft, find the work done by the force in ft-lb.

5. If $\mathbf{a} = l\mathbf{i} + m\mathbf{j} + n\mathbf{k}$, $\mathbf{b} = l'\mathbf{i} + m'\mathbf{j} + n'\mathbf{k}$ are two unit vectors and θ is the angle between them, show that $\cos \theta = ll' + mm' + nn'$.

6. Find the scalar product of the vectors represented by two diagonals of a unit cube, and find also the angle between these diagonals.

22   CONCISE VECTOR ANALYSIS

2.2. Vector product

From two vectors **a** and **b** a product which is a vector may be formed in the following way:

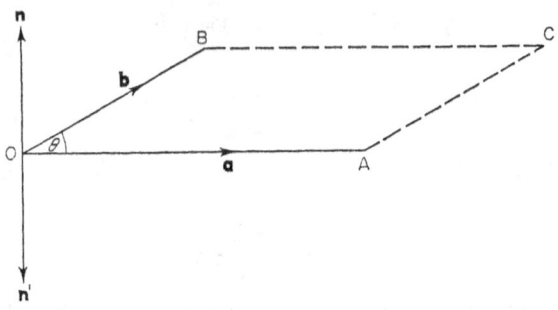

Fig. 5

Suppose $\mathbf{a} = \overrightarrow{OA}$ and $\mathbf{b} = \overrightarrow{OB}$ are the two vectors, and θ is the angle, less than two right angles, through which OA must be turned in the plane AOB to coincide with OB. Let us draw through O the normal to the plane AOB. This can be drawn in two senses, which are opposite to one another. Let **n** be a unit vector along the normal in such a sense that **a, b, n**, taken in that order form a right-hand system, that is, if a right screw is placed with its axis along **n** and the screw is turned about its axis so that OA turns towards OB, then the screw will move forward along **n**. **n**′ is the unit normal to the plane AOB in the sense opposite to **n**, that is, $\mathbf{n}' = -\mathbf{n}$.

The *vector product* of **a** and **b**, taken in that order (also sometimes called *outer* product) is defined to be the vector with the following properties:

(i) its magnitude is $|\mathbf{a}||\mathbf{b}| \sin \theta$
(ii) its direction is along the unit vector **n** which is perpendicular to **a** and also to **b** and hence is normal to the plane formed by **a** and **b**, and such that **a, b, n** taken in that order form a right-hand system.

PRODUCTS OF VECTORS 23

The vector product so defined is denoted by **a** × **b**, and read "**a** cross **b**". (It is also sometimes written **a** ∧ **b**, read "**a** vec **b**.") With the above notation

$$\mathbf{a} \times \mathbf{b} = |\mathbf{a}||\mathbf{b}| \sin \theta \, \mathbf{n}. \tag{16}$$

Now let us consider the vector product **b** × **a**. According to the definition its magnitude is $|\mathbf{b}||\mathbf{a}| \sin \theta$ which is the same as the magnitude of **a** × **b**. The direction of **b** × **a** is along **n**′, since **b**, **a**, **n**′ taken in that order form a right-hand system. Hence

$$\mathbf{b} \times \mathbf{a} = |\mathbf{b}||\mathbf{a}| \sin \theta \, \mathbf{n}' = -|\mathbf{b}||\mathbf{a}| \sin \theta \, \mathbf{n} = -\mathbf{a} \times \mathbf{b}. \tag{17}$$

Thus we see that the commutative law does not hold for vector product. We have therefore to be careful of the order in which we write the factors in a vector product.

It is easy to verify the following properties:

$$\mathbf{a} \times \mathbf{a} = \mathbf{O}. \tag{18}$$

If **a**, **b** are parallel vectors **a** × **b** = **O**.
If λ is a scalar

$$(\lambda \mathbf{a}) \times \mathbf{b} = \lambda \mathbf{a} \times \mathbf{b} = \mathbf{a} \times (\lambda \mathbf{b}). \tag{19}$$

If
$$\mathbf{a} \times \mathbf{b} = \mathbf{O} \tag{20}$$

either **a** = **O** or **b** = **O** or **a**, **b** are parallel or in the same line.

Example. If **i**, **j**, **k** are unit vectors along the three axes of coordinates, find **i** × **i**, **i** × **j**, **i** × **k**, **j** × **i**, **j** × **j**, **j** × **k**, **k** × **i**, **k** × **j**, **k** × **k**.

We see that

$$\mathbf{i} \times \mathbf{i} = \mathbf{j} \times \mathbf{j} = \mathbf{k} \times \mathbf{k} = \mathbf{O}, \tag{21}$$

being particular cases of (18). Also, **i**, **j**, **k**, in this order, form a right-hand system. So do those vectors in the order obtained by cyclic interchange of **i, j, k**, that is, **j, k, i** and **k, i, j**. Hence we see that

$$\left.\begin{array}{l}\mathbf{i} \times \mathbf{j} = \mathbf{k} = -\mathbf{j} \times \mathbf{i} \\ \mathbf{j} \times \mathbf{k} = \mathbf{i} = -\mathbf{k} \times \mathbf{j} \\ \mathbf{k} \times \mathbf{i} = \mathbf{j} = -\mathbf{i} \times \mathbf{k} \end{array}\right\} \quad (22)$$

Among other properties of vector products which are of use is the result that $\mathbf{a} \times \mathbf{b}$ is not altered if we replace one factor by its orthogonal projection on a plane perpendicular to the other. Let B' be the foot of the perpendicular from B on to the plane through O to which OA is normal (Fig. 6). Then OA, OB, and OB' all lie on a plane to which $\mathbf{a} \times \mathbf{b}$ is normal. Therefore

$$OB' = OB \cos(\pi/2 - \theta) = OB \sin \theta.$$

If $\overrightarrow{OB'} \equiv \mathbf{b}'$, $|\mathbf{b}'| = |\mathbf{b}| \sin \theta$, and $\mathbf{a} \times \mathbf{b}' = |\mathbf{a}||\mathbf{b}'|\mathbf{n}$

$$= |\mathbf{a}||\mathbf{b}| \sin \theta \, \mathbf{n} = \mathbf{a} \times \mathbf{b}. \quad (23)$$

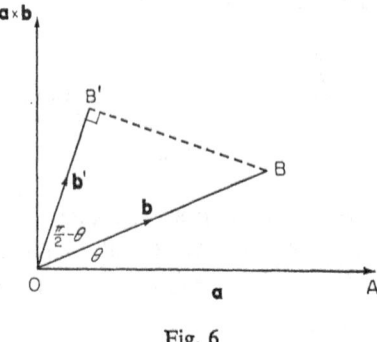

Fig. 6

This result is useful in establishing the distributive law for vector products, that is,

$$\mathbf{a} \times (\mathbf{b} + \mathbf{c}) = \mathbf{a} \times \mathbf{b} + \mathbf{a} \times \mathbf{c}. \quad (24)$$

PRODUCTS OF VECTORS

To prove this result, suppose we take the plane of the paper to be a plane perpendicular to the vector **a**, and let **b**′, **c**′ and (**b** + **c**)′ be the projections of **b**, **c** and (**b** + **c**) respectively on this plane. By application of (23) we have **a** × **b** = **a** × **b**′ and **a** × **c** = **a** × **c**′. Therefore to prove (24) it is enough to show that

$$\mathbf{a} \times (\mathbf{b} + \mathbf{c})' = \mathbf{a} \times \mathbf{b}' + \mathbf{a} \times \mathbf{c}'. \tag{25}$$

Let the triangle PQR in the figure have sides representing **b**′, **c**′ and (**b** + **c**)′. We see that (**b** + **c**)′ = **b**′ + **c**′. Let $P'Q'$ represent **a** × **b**′ and $Q'R'$ represent **a** × **c**′.

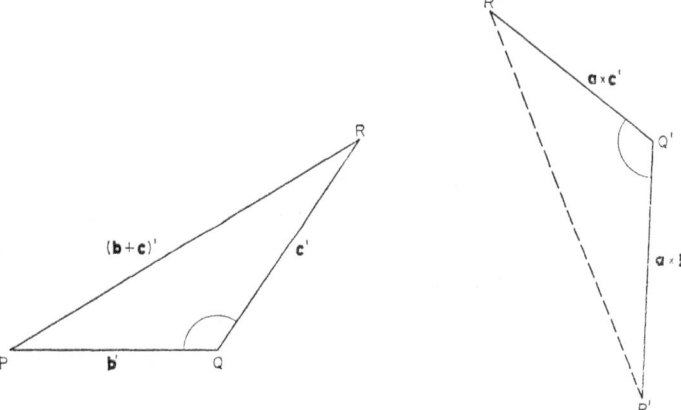

Fig. 7

We see that in the figure $P'Q'$ and $Q'R'$ are perpendicular to and proportional to PQ and QR respectively. Therefore the triangles PQR and $P'Q'R'$ are similar. The triangle $P'Q'R'$ may be obtained from the triangle PQR by turning it in its plane through a right angle, and magnifying each side in the ratio $|\mathbf{a}|:1$. Hence the third side $P'R'$ has magnitude $|\mathbf{a}| \cdot PR = |\mathbf{a}| \, |(\mathbf{b} + \mathbf{c})'|$ and represents $\mathbf{a} \times (\mathbf{b} + \mathbf{c})'$. Since $\overrightarrow{P'R'} = \overrightarrow{P'Q'} + \overrightarrow{Q'R'}$, we obtain (25) and hence (24).

In terms of Cartesian components of **a** and **b**, **a** × **b** may be obtained as follows:

$$\begin{aligned}\mathbf{a} \times \mathbf{b} &= (a_x\mathbf{i} + a_y\mathbf{j} + a_z\mathbf{k}) \times (b_x\mathbf{i} + b_y\mathbf{j} + b_z\mathbf{k}) \\ &= a_xb_x\mathbf{i} \times \mathbf{i} + a_xb_y\mathbf{i} \times \mathbf{j} + a_xb_z\mathbf{i} \times \mathbf{k} \\ &\quad + a_yb_x\mathbf{j} \times \mathbf{i} + a_yb_y\mathbf{j} \times \mathbf{j} + a_yb_z\mathbf{j} \times \mathbf{k} + a_zb_x\mathbf{k} \times \mathbf{i} \\ &\quad + a_zb_y\mathbf{k} \times \mathbf{j} + a_zb_z\mathbf{k} \times \mathbf{k} \\ &= \mathbf{i}(a_yb_z - a_zb_y) + \mathbf{j}(a_zb_x - a_xb_z) + \mathbf{k}(a_xb_y - a_yb_x). \quad (26)\end{aligned}$$

This may be written in the form

$$\mathbf{a} \times \mathbf{b} = \begin{vmatrix} \mathbf{i} & \mathbf{j} & \mathbf{k} \\ a_x & a_y & a_z \\ b_x & b_y & b_z \end{vmatrix}. \quad (27)$$

These various properties of vector products enable us to manipulate expressions and equations containing vector products. The rules of ordinary algebra may be followed, with the provision that the order of factors in vector products has to be carefully maintained. For example, we may use the above to infer that

$$(\mathbf{a} + \mathbf{b}) \times (\mathbf{c} + \mathbf{d}) = \mathbf{a} \times \mathbf{c} + \mathbf{a} \times \mathbf{d} + \mathbf{b} \times \mathbf{c} + \mathbf{b} \times \mathbf{d}. \quad (28)$$

Illustrations

1. In Fig. 5, the area of the parallelogram is

$$|\mathbf{a}||\mathbf{b}| \sin \theta = |\mathbf{a} \times \mathbf{b}|.$$

Also the normal to the plane of the parallelogram is along the direction of **n**, that is, along the same direction as the vector **a** × **b**. If we adopt the convention that an area A may be represented vectorially by a vector whose direction is normal to the area, and whose magnitude is proportional to the magnitude of the area, we see that the vector product **a** × **b** represents vectorially the area of the parallelogram $OACB$ which has adjacent sides representing the vectors **a** and **b** respectively.

PRODUCTS OF VECTORS

2. *Moment of a localized vector about a point*:

If **F** is a vector localized along a line, and **r** the position vector of a point R on the line of action of **F**, the vector moment about O of the vector **F** is defined to be the vector **M** where

$$\mathbf{M} = \mathbf{r} \times \mathbf{F}. \tag{29}$$

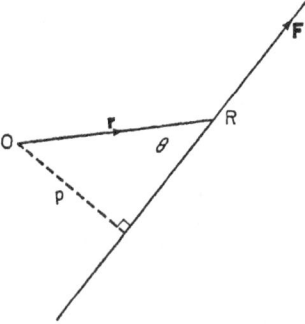

Fig. 8

This moment is independent of the particular position of R on the line. If we take another point R' on the line, and its position vector is $\mathbf{r}'$, then

$$\mathbf{r}' = \mathbf{r} + \overrightarrow{RR'}$$

and

$$\mathbf{r}' \times \mathbf{F} = (\mathbf{r} + \overrightarrow{RR'}) \times \mathbf{F} = \mathbf{r} \times \mathbf{F} + \overrightarrow{RR'} \times \mathbf{F} = \mathbf{r} \times \mathbf{F},$$

since $\overrightarrow{RR'} \times \mathbf{F} = \mathbf{O}$, as both RR' and **F** are along the same line. The magnitude of the moment **M** is $|\mathbf{r}||\mathbf{F}| \sin \theta = p|\mathbf{F}|$ where p is the perpendicular distance of the line of action of **F** from O. The direction of **M** is normal to the plane containing the point O and the line of action of **F**.

The moment of **F** about a point whose position vector is $\mathbf{r}_0$ is

$$(\mathbf{r} - \mathbf{r}_0) \times \mathbf{F}. \tag{30}$$

Exercise IIb

1. If $\mathbf{a} = 2\mathbf{j} - \mathbf{k}$, $\mathbf{b} = \mathbf{i} - 2\mathbf{j}$, $\mathbf{c} = \mathbf{i} + \mathbf{j} + \mathbf{k}$, evaluate
    - (i) $\mathbf{a} \times \mathbf{b}$
    - (ii) $\mathbf{b} \times \mathbf{c}$
    - (iii) $(\mathbf{a} \times \mathbf{b}) \cdot \mathbf{c}$
    - (iv) $\mathbf{a} \cdot (\mathbf{b} \times \mathbf{c})$
    - (v) $\mathbf{a} \times (\mathbf{b} \times \mathbf{c})$
    - (vi) $\mathbf{a} \cdot (\mathbf{a} \times \mathbf{b})$
    - (vii) $\mathbf{a} \times (\mathbf{a} \times \mathbf{b})$.

2. If the vertices A, B, C of a triangle have position vectors $\mathbf{i}$, $\mathbf{i} + \mathbf{j} - \mathbf{k}$ and $-\mathbf{k}$ respectively, find
    - (i) the values of sin A, sin B, sin C
    - (ii) unit vector normal to the plane of the triangle
    - (iii) the area of the triangle.

3. The vector (1, 2, 5) is drawn from the origin. Find the vector drawn perpendicular to this from the point (1, 0, 1).

4. The end points of three vectors $\mathbf{a}$, $\mathbf{b}$, $\mathbf{c}$, drawn from the origin O of coordinates, are the vertices of a triangle ABC.
    Find (i) the distance from the origin of the plane of the triangle
    (ii) the area of the triangle
    (iii) the volume of the tetrahedron $OABC$.

5. If $\mathbf{F} = (X, Y, 0)$ and $\mathbf{r} = (x, y, 0)$ show that the vector moment about the origin of the vector $\mathbf{F}$ acting at the point $\mathbf{r}$ is $(0, 0, xY - yX)$.

2.3. Triple products

If $\mathbf{a}$, $\mathbf{b}$, $\mathbf{c}$ are any three vectors, the scalar product of $\mathbf{a} \times \mathbf{b}$ with $\mathbf{c}$ is written as $(\mathbf{a} \times \mathbf{b}) \cdot \mathbf{c}$ and is known as a scalar triple product.

We can see that this product is numerically equal to the volume of the parallelepiped which has $\mathbf{a}$, $\mathbf{b}$, $\mathbf{c}$ as adjacent edges. Consider the case when the angle between $\mathbf{a} \times \mathbf{b}$ and $\mathbf{c}$ is an acute angle ϕ.

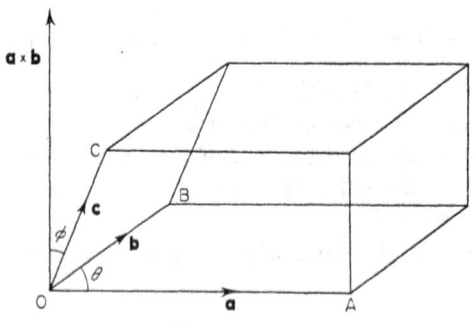

Fig. 9

PRODUCTS OF VECTORS

$$(\mathbf{a} \times \mathbf{b}).\mathbf{c} = |\mathbf{a} \times \mathbf{b}||\mathbf{c}| \cos \phi$$
$$= |\mathbf{a}_1||\mathbf{b}| \sin \theta |\mathbf{c}| \cos \phi.$$

If we consider the volume of the parallelepiped with edges OA, OB, OC, we may take the face containing OA and OB as the base.

The area of the base is $|\mathbf{a}| |\mathbf{b}| \sin \theta$, and the altitude of the parallelepiped is $|\mathbf{c}| \cos \phi$. Hence we obtain that the volume of the parallelepiped is $|\mathbf{a}| |\mathbf{b}| \sin \theta |\mathbf{c}| \cos \phi$ which is also the scalar triple product $(\mathbf{a} \times \mathbf{b}).\mathbf{c}$.

In the general case when the angle ϕ need not be acute,

$$(\mathbf{a} \times \mathbf{b}).\mathbf{c} = \pm \text{volume of parallelepiped}. \tag{31}$$

We can show that

$$(\mathbf{a} \times \mathbf{b}).\mathbf{c} = \mathbf{a}.(\mathbf{b} \times \mathbf{c}). \tag{32}$$

We may denote each side by the symbol $[\mathbf{a}, \mathbf{b}, \mathbf{c},]$ or by $(\mathbf{a}\,\mathbf{b}\,\mathbf{c})$ or $(\mathbf{a}, \mathbf{b}, \mathbf{c})$.

Using eqn. (27), we can obtain

$$[\mathbf{a}, \mathbf{b}, \mathbf{c}] = \begin{vmatrix} a_x & a_y & a_z \\ b_x & b_y & b_z \\ c_x & c_y & c_z \end{vmatrix}. \tag{33}$$

Using properties of determinants, we may see that

$$[\mathbf{a}, \mathbf{b}, \mathbf{c}] = [\mathbf{b}, \mathbf{c}, \mathbf{a}] = [\mathbf{c}, \mathbf{a}, \mathbf{b}]$$
$$= -[\mathbf{a}, \mathbf{c}, \mathbf{b}] = -[\mathbf{b}, \mathbf{a}, \mathbf{c}] = -[\mathbf{c}, \mathbf{b}, \mathbf{a}]. \tag{34}$$

We see that those products in which the order of three vectors can be obtained from $\mathbf{a}$, $\mathbf{b}$, $\mathbf{c}$ by cyclic interchange have one sign, while those which come from cyclic interchange of $\mathbf{a}$, $\mathbf{c}$, $\mathbf{b}$ have the opposite sign.

If $\mathbf{a}$, $\mathbf{b}$, $\mathbf{c}$ are coplanar vectors then $[\mathbf{a}, \mathbf{b}, \mathbf{c}] = 0$.

If

$$[\mathbf{a}, \mathbf{b}, \mathbf{c}] = 0,$$

we can infer that either $\mathbf{a} = \mathbf{O}$ or $\mathbf{b} = \mathbf{O}$ or $\mathbf{c} = \mathbf{O}$ or $\mathbf{a}$, $\mathbf{b}$, $\mathbf{c}$ are coplanar.

In particular [**a**, **b**, **c**] = 0 if any two of **a**, **b**, **c** are the same, e.g. [**a**, **a**, **b**] = 0 for any **a** and **b**.

A **vector triple product**, such as (**a** × **b**) × **c**, may also be defined. The position of the bracket is important in this case since it would be seen that (**a** × **b**) × **c** is different from **a** × (**b** × **c**). Using formula (26) it may be shown that

$$\mathbf{a} \times (\mathbf{b} \times \mathbf{c}) = (\mathbf{a} \cdot \mathbf{c})\mathbf{b} - (\mathbf{a} \cdot \mathbf{b})\mathbf{c}. \tag{35}$$

2.4. Mutual moment of two lines

Suppose R, R' are two points whose position vectors are **r** and **r**′, and which lie on the lines l and l' respectively. Let **u**, **u**′ be unit vectors along the two lines. The vector moment **M** about R of a vector **F** along l' is given by

$$\mathbf{M} = (\mathbf{r}' - \mathbf{r}) \times F\mathbf{u}'.$$

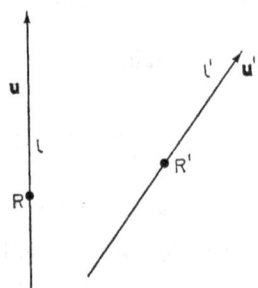

Fig. 10

The scalar moment about the line l of the vector **F** along l' is defined as the component along l of the vector moment about the point R, that is, $\text{comp}_l \mathbf{M}$. This has the value

$$\mathbf{M} \cdot \mathbf{u} = F[\mathbf{r}' - \mathbf{r}, \mathbf{u}', \mathbf{u}]. \tag{36}$$

We can verify that R can be any point on l and R' any point on l', without the expression (36) being altered.

We may also note that the moment about one line of unit vector along the second line is equal to the moment about the

PRODUCTS OF VECTORS   31

second line of unit vector along the first line. We may establish this result by noting that $[\mathbf{r} - \mathbf{r}', \mathbf{u}, \mathbf{u}'] = [\mathbf{r}' - \mathbf{r}, \mathbf{u}', \mathbf{u},]$ which equality may be seen from the properties of determinants.

We define

$$m = [\mathbf{r} - \mathbf{r}', \mathbf{u}, \mathbf{u}'] = [\mathbf{r}' - \mathbf{r}, \mathbf{u}', \mathbf{u}] \tag{37}$$

as the mutual moment of the two lines. If the two lines intersect $m = 0$.

2.5. Positive and negative triads

If $\mathbf{i}, \mathbf{j}, \mathbf{k}$ denote unit vectors along the axes of a right-handed system of coordinates, we see that the triple scalar product $(\mathbf{i}, \mathbf{j}, \mathbf{k}) = +1$. We say in this case that the three unit vectors $\mathbf{i}, \mathbf{j}, \mathbf{k}$ in this order form a positive triad. It is easy to see that $\mathbf{j}, \mathbf{k}, \mathbf{i}$ and $\mathbf{k}, \mathbf{i}, \mathbf{j}$ (which are obtained from $\mathbf{i}, \mathbf{j}, \mathbf{k}$ by cyclic interchange) are also positive triads. But $\mathbf{i}, \mathbf{k}, \mathbf{j}$ and $\mathbf{j}, \mathbf{i}, \mathbf{k}$ and $\mathbf{k}, \mathbf{j}, \mathbf{i}$ belong to left-handed systems and are negative triads, the scalar products in these orders having value -1. Triads with the same cyclic order have the same signs.

This notion can be extended to any triad of non-coplanar vectors $\mathbf{a}, \mathbf{b}, \mathbf{c}$. It is possible to deform such a triad into a triad of three mutually perpendicular vectors by rotating the vectors through acute angles without any one of the vectors crossing the plane of the other two. We say that $\mathbf{a}, \mathbf{b}, \mathbf{c}$ form a positive or negative triad according as the three mutually perpendicular vectors, into which the triad $\mathbf{a}, \mathbf{b}, \mathbf{c}$ may be deformed, form a positive or negative triad.

When $\mathbf{a}, \mathbf{b}, \mathbf{c}$ form a positive triad the triple scalar product $[\mathbf{a}, \mathbf{b}, \mathbf{c}]$ has a positive value. For a negative triad the triple scalar product is negative. If $\mathbf{a}, \mathbf{b}, \mathbf{c}$ form a positive triad, so do $\mathbf{b}, \mathbf{c}, \mathbf{a}$ and $\mathbf{c}, \mathbf{a}, \mathbf{b}$. But $\mathbf{a}, \mathbf{c}, \mathbf{b}$ and $\mathbf{b}, \mathbf{a}, \mathbf{c}$ and $\mathbf{c}, \mathbf{b}, \mathbf{a}$ will then be negative triads. We note in the definition of vector products given earlier, the sense of $\mathbf{a} \times \mathbf{b}$ was taken in such a way that $\mathbf{a}, \mathbf{b}$ and $\mathbf{a} \times \mathbf{b}$, in that order, form a positive triad.

Exercise IIc

1. Given $\mathbf{a} = \mathbf{i} - \mathbf{j}$, $\mathbf{b} = \mathbf{i} - 2\mathbf{j} + \mathbf{k}$, $\mathbf{c} = \mathbf{i} + 3\mathbf{j}$, determine whether $\mathbf{a}$, $\mathbf{b}$, $\mathbf{c}$ form a positive or negative triad. Evaluate
   (*i*) $\mathbf{a} \cdot [(\mathbf{a} - \mathbf{b}) \times \mathbf{c}]$ (*ii*) $(\mathbf{a} \times \mathbf{c}) \cdot (\mathbf{b} \times \mathbf{c})$ (*iii*) $\mathbf{a} \times (\mathbf{b} \times \mathbf{c})$
   (*iv*) $(\mathbf{a} \cdot \mathbf{c})\mathbf{b} - (\mathbf{a} \cdot \mathbf{b})\mathbf{c}$.

2. Show that
$$\mathbf{a} \times (\mathbf{b} \times \mathbf{c}) + \mathbf{b} \times (\mathbf{c} \times \mathbf{a}) + \mathbf{c} \times (\mathbf{a} \times \mathbf{b}) = 0.$$

3. Find the vector moment about the point $\mathbf{i} - 2\mathbf{j} + \mathbf{k}$ of a force represented by $3\mathbf{i} + \mathbf{k}$ acting through the point $2\mathbf{i} - \mathbf{j} + 3\mathbf{k}$.

4. A vector $\mathbf{x}$ is written in the form
$$\mathbf{x} = \lambda\mathbf{a} + \mu\mathbf{b} + \nu\mathbf{c}$$
where $\mathbf{a}$, $\mathbf{b}$, $\mathbf{c}$ are three given non-coplanar vectors. Show that
$$\lambda = \frac{[\mathbf{x}, \mathbf{b}, \mathbf{c}]}{[\mathbf{a}, \mathbf{b}, \mathbf{c}]}, \quad \mu = \frac{[\mathbf{x}, \mathbf{c}, \mathbf{a}]}{[\mathbf{b}, \mathbf{c}, \mathbf{a}]}, \quad \nu = \frac{[\mathbf{x}, \mathbf{a}, \mathbf{b}]}{[\mathbf{c}, \mathbf{a}, \mathbf{b}]}.$$
Interpret the result when $\mathbf{a}$, $\mathbf{b}$, $\mathbf{c}$ are taken to be $\mathbf{i}$, $\mathbf{j}$, $\mathbf{k}$.

5. If θ is the angle between the two vectors $\mathbf{a} = 2\mathbf{j} - \mathbf{k}$, $\mathbf{b} = \mathbf{i} + 2\mathbf{j}$, determine (*i*) a unit vector that is perpendicular to both $\mathbf{a}$ and $\mathbf{b}$,
   (*ii*) $\sin \theta$.

6. $A = (3, 7, -2)$, $B = (1, 5, -1)$, $\overrightarrow{AC} = \mathbf{i} + 2\mathbf{j}$, $\overrightarrow{BD} = -\mathbf{i} + \mathbf{k}$, show that the lines AC and BD intersect.

7. If θ is the angle between two skew lines and d the shortest distance between them, show that the mutual moment $m = d \sin \theta$.

8. The unit vectors $\mathbf{a}$ and $\mathbf{b}$ are perpendicular and the unit vector $\mathbf{c}$ is inclined at an angle θ to both a and b. Show that
$$\mathbf{c} = \alpha(\mathbf{a} + \mathbf{b}) + \beta(\mathbf{a} \times \mathbf{b}),$$
where
$$\alpha = \cos \theta \quad \text{and} \quad \beta^2 = -\cos 2\theta.$$

The vector $\mathbf{x}$ satisfies the equation
$$\mathbf{x} = \mathbf{a} \times \mathbf{b} + \mathbf{c} \times \mathbf{x}.$$
Show that
$$2\mathbf{x} = \mathbf{a} \times \mathbf{b} + \mathbf{c} \times (\mathbf{a} \times \mathbf{b}) + (\mathbf{c} \cdot \mathbf{x})\mathbf{c}$$
and hence obtain the solution
$$\mathbf{x} = \tfrac{1}{2}\alpha\left(1 - \frac{1}{\beta}\right)\mathbf{a} - \tfrac{1}{2}\alpha\left(1 + \frac{1}{\beta}\right)\mathbf{b} + \frac{1 - \alpha^2}{\beta}\mathbf{c}.$$

[Camb. N.S. 1957]

PRODUCTS OF VECTORS   33

9. Vectors **a**, **b**, **c** are such that **a** is perpendicular both to **b** and to **c**, and $|\mathbf{b}| = |\mathbf{c}|$. Show that the equation of the plane through the three points whose position vectors are **a**, **b**, **c** is

$$\left(\frac{\mathbf{a}}{|\mathbf{a}|^2} + \frac{\mathbf{b}+\mathbf{c}}{|\mathbf{b}||\mathbf{c}| + \mathbf{b}.\mathbf{c}}\right).\mathbf{r} = 1.$$

Hence, or otherwise, find the equation of the plane through the points

$$(1, 1, 1), \quad (-1, 2, -1), \quad (-1, -1, 2).$$

[Camb. N.S. 1956]

10. Three vectors are written in the form

$$\mathbf{r}_n = x_n\mathbf{i} + y_n\mathbf{j} + z_n\mathbf{k} \quad (n = 1, 2, 3),$$

where **i**, **j**, **k** are three non-orthogonal (but not coplanar) unit vectors. Show that the numbers x_n can be written in the form

$$x_n = \frac{\mathbf{r}_n.(\mathbf{j}\times\mathbf{k})}{\mathbf{i}.(\mathbf{j}\times\mathbf{k})},$$

and give two similar expressions for y_n and z_n.

By considering the vector product of $\mathbf{r}_1$ and $\mathbf{r}_2$, or otherwise, show that $\mathbf{r}_1$, $\mathbf{r}_2$ and $\mathbf{r}_3$ will be coplanar if

$$\begin{vmatrix} x_1 & y_1 & z_1 \\ x_2 & y_2 & z_2 \\ x_3 & y_3 & z_3 \end{vmatrix} = 0.$$

[Camb. P.N.S. 1957]

11. Three planes have the equations $\mathbf{r}.\mathbf{l} = 0$, $\mathbf{r}.\mathbf{m} = 0$, $\mathbf{r}.\mathbf{n} = 0$, where **l**, **m** and **n** are unit vectors. Give the conditions that the vector **p** should be equally inclined to these unit vectors. Find such a vector **p** for the case where **l**, **m** and **n** point in the directions (1, 2, 2), (2, 3, 6) and (0, 3, 4) respectively. Deduce that there exists a cone of semi-vertical angle $\cos^{-1}(1/\sqrt{(26)})$ which touches all three planes, and give the vector equation of this cone.

[Camb. P.N.S. 1956]

12. The vector equations of 2 lines are

$$\mathbf{x} = \mathbf{a} + r\mathbf{l}, \quad \mathbf{x} = \mathbf{b} + s\mathbf{m}$$

where **l**, **m** are unit vectors and r, s are scalar parameters.
Prove that the length of the shortest distance between them is

$$(\mathbf{a} - \mathbf{b}, \mathbf{l}, \mathbf{m}) \operatorname{cosec} \theta$$

where $\cos\theta = \mathbf{l}.\mathbf{m}$.

Find the shortest distance PP' between the lines whose equations in rectangular Cartesian coordinates are $x - c = y + c = 6z$, $x = 2(y - c) = -12(z + c)$.
Find also the coordinates of the midpoint of PP'.

[Camb. P.M. 1945]

13. (*i*) Solve the following equation for λ, μ, ν in terms of the three-dimensional vectors **a**, **b**, **c**, **d**:

$$\lambda \mathbf{a} + \mu \mathbf{b} + \nu \mathbf{c} = \mathbf{d}.$$

(*ii*) Show that if two pairs of opposite sides of a tetrahedron are perpendicular, then the third pair are also pependicular.

[Camb. N.S. 1953]

14. *P* is the foot of the perpendicular from a point *B*, with position vector **b**, to the line $\mathbf{r} = \mathbf{a} + \lambda \mathbf{t}$. Show that the equation of the line *BP* is

$$\mathbf{r} = \mathbf{b} + \mu \mathbf{t} \times \{(\mathbf{a} - \mathbf{b}) \times \mathbf{t}\},$$

and find the position vector of *P*.

[Camb. P.N.S. 1953]

15. Define the vector product $\mathbf{a} \times \mathbf{b}$ of two vectors **a** and **b**, and discuss its direction and magnitude in relation to those of **a** and **b**.
Prove the formulae:

(*i*) $\mathbf{a} \times (\mathbf{b} \times \mathbf{c}) = (\mathbf{a}.\mathbf{c})\mathbf{b} - (\mathbf{a}.\mathbf{b})\mathbf{c}$,

(*ii*) $\mathbf{a} \times (\mathbf{b} \times \mathbf{c}) + \mathbf{b} \times (\mathbf{c} \times \mathbf{a}) + \mathbf{c} \times (\mathbf{a} \times \mathbf{b}) = \mathbf{O}$.

Show further, by use of (*ii*) or otherwise, that if two trihedrals *OABC* and *OA'B'C'*, with the same vertex *O*, are such that *OA*, *OB*, *OC* are perpendicular to the planes *OB'C'*, *OC'A'*, *OA'B'* respectively, then the planes *OAA'*, *OBB'* and *OCC'* have a line in common.

[Lond. G.II 1959]

16. Explain what is meant by asserting the linear independence of a set of vectors. If **a**, **b**, **c** are three linearly independent space vectors, prove that any other space vector **d** can be expressed in the form $\mathbf{d} = p\mathbf{a} + q\mathbf{b} + r\mathbf{c}$, where p, q, r are scalars.

If **a**, **b**, **c**, **d** are the position vectors of the vertices of a tetrahedron *ABCD*, prove that

(*i*) the vector $\mathbf{b} \times \mathbf{c} + \mathbf{c} \times \mathbf{a} + \mathbf{a} \times \mathbf{b}$ is perpendicular to the plane *ABC* and of magnitude equal to twice the area of the triangle *ABC*,

(*ii*) the equation of the plane *ABC* is

$$\mathbf{x}.(\mathbf{b} \times \mathbf{c} + \mathbf{c} \times \mathbf{a} + \mathbf{a} \times \mathbf{b}) = (\mathbf{abc}),$$

(*iii*) the volume of the tetrahedron *ABCD* is

$$\pm \tfrac{1}{6} \{(\mathbf{abc}) - (\mathbf{bcd}) - (\mathbf{cad}) - (\mathbf{abd})\}.$$

[Lond. G.II 1960]

17. If $[\mathbf{a}, \mathbf{b}, \mathbf{c}] = \mathbf{a}.(\mathbf{b} \times \mathbf{c})$ show by geometrical considerations that

$$[\mathbf{a}, \mathbf{b}, \mathbf{c}] = [\mathbf{b}, \mathbf{c}, \mathbf{a}] = -[\mathbf{b}, \mathbf{a}, \mathbf{c}].$$

If $\mathbf{R} = l\mathbf{a} + m\mathbf{b} + n\mathbf{c}$, $\mathbf{G} = \mathbf{c} \times m\mathbf{b} + (\mathbf{a} - \mathbf{b}) \times n\mathbf{c}$ prove that

$$\mathbf{R}.\mathbf{G} = -(mn + nl + lm)[\mathbf{a}, \mathbf{b}, \mathbf{c}].$$

[Camb. N.S.I. 1943]

PRODUCTS OF VECTORS   35

18. **a, b, c** are three coplanar vectors; **b, c** are perpendicular to one another and have magnitudes b, c respectively. If **r** is the radius vector from some origin and λ a scalar parameter taking all values from $-\infty$ to ∞, show that $\mathbf{r} = \mathbf{a} + \lambda\mathbf{b} + \lambda^2\mathbf{c}$ represents a parabola of latus rectum $\mathbf{b}^2/c$ and focus $\mathbf{a} + (b^2/4c^2)\mathbf{c}$. Show also that at any point λ on the parabola the tangent is in the direction of the vector $\mathbf{b} + 2\lambda\mathbf{c}$ and the normal is in the direction of the vector $2\lambda c^2\mathbf{b} - b^2\mathbf{c}$.

[Camb. P.N.S. 1955]

19. The position vectors of three points relative to a given origin are **P, Q, R**. Show that a necessary and sufficient condition for these vectors to be coplanar is that $\mathbf{P}\cdot(\mathbf{Q} \times \mathbf{R})$ should vanish.

The position vectors of three distinct points on a sphere of unit radius and centre the origin are **A, B, C** where $\mathbf{A}\cdot(\mathbf{B} \times \mathbf{C}) \neq 0$. The angle between **B** and **C**, **C** and **A**, **A** and **B** are a, b, c respectively. A fourth point has position vector **D** such that

$$\mathbf{D}\cdot\mathbf{A} = \alpha, \quad \mathbf{D}\cdot\mathbf{B} = \beta, \quad \mathbf{D}\cdot\mathbf{C} = 0.$$

Show that the magnitude of its components perpendicular to the plane of **A** and **B** is

$$\left|\frac{\alpha(\cos c \cos a - \cos b) + \beta(\cos b \cos c - \cos a)}{(\mathbf{A}\cdot\mathbf{B} \times \mathbf{C}) \sin c}\right|.$$

[Camb. M.T. 1959]

20. Define the vector product $\mathbf{u} \times \mathbf{v}$, and show that if **u** is of unit magnitude then $\{\mathbf{u} \times \mathbf{v}\} \times \mathbf{u}$ is the component of **v** perpendicular to **u**.

A and B are fixed points and **w** is a given vector perpendicular to AB. Describe in geometrical terms the loci defined by the following vector equations in three-dimensional space:

$$(i)\ \overrightarrow{AP}\cdot\overrightarrow{BP} = 0, \quad (ii)\ \overrightarrow{AP} \times \overrightarrow{BP} = \mathbf{w}.$$

[Lond. G.II, 1957]

21. If **a, b, u** are any three vectors, show that

$$(\mathbf{a} \times \mathbf{b}) \times \mathbf{u} = (\mathbf{a}\cdot\mathbf{u})\mathbf{b} - (\mathbf{b}\cdot\mathbf{u})\mathbf{a}.$$

By considering $(\mathbf{a} \times \mathbf{b}) \times (\mathbf{c} \times \mathbf{d})$, or otherwise, show that

$$[\mathbf{a}, \mathbf{b}, \mathbf{c}]\mathbf{d} = [\mathbf{b}, \mathbf{c}, \mathbf{d}]\mathbf{a} + [\mathbf{c}, \mathbf{a}, \mathbf{d}]\mathbf{b} + [\mathbf{a}, \mathbf{b}, \mathbf{d}]\mathbf{c},$$

where [**a, b, c**] etc. are scalar triple products. Evaluate the triple products on the right when $\mathbf{d} = \mathbf{a} \times \mathbf{b}$, expressing them in terms of scalar products of **a, b** and **c**.

[Camb. P.M. 1960]

22. Find the vector from O to the common point of the three planes given respectively by $\mathbf{r}\cdot\mathbf{A} = 1$, $\mathbf{r}\cdot\mathbf{B} = 1$ and $\mathbf{r}\cdot\mathbf{C} = 1$ where **A, B** and **C** are three non-coplanar const. vectors and **r** is the radius vector from O. If **a, b, c,** and **d** are four vectors drawn from O, show that the point equidistant from their end points is given by

$$\mathbf{r} = -\frac{1}{2}\frac{\mathbf{a}^2\{\mathbf{b}, \mathbf{c}, \mathbf{d}\} + \mathbf{b}^2\{\mathbf{c}, \mathbf{a}, \mathbf{d}\} + \mathbf{c}^2\{\mathbf{a}, \mathbf{b}, \mathbf{d}\} - \mathbf{d}^2\{\mathbf{a}, \mathbf{b}, \mathbf{c}\}}{[\mathbf{b}, \mathbf{c}, \mathbf{d}] + [\mathbf{c}, \mathbf{a}, \mathbf{d}] + [\mathbf{a}, \mathbf{b}, \mathbf{d}] - [\mathbf{a}, \mathbf{b}, \mathbf{c}]},$$

where $\{a, b, c\} \equiv a \times b + b \times c + c \times a$
$[a, b, c] \equiv a.(b \times c).$

[Camb. P.M. 1948]

23. The set of vectors b_1, b_2, b_3 reciprocal to the set a_1, a_2, a_3 is defined by

$$b_1 = \frac{a_2 \times a_3}{v_a}, \quad b_2 = \frac{a_3 \times a_1}{v_a}, \quad b_3 = \frac{a_1 \times a_2}{v_a}$$

where
$$v_a = a_1.a_2 \times a_3.$$

Show that
$$b_1.b_2 \times b_3 = \frac{1}{a_1.a_2 \times a_3}.$$

Axes parallel to a_1, a_2, a_3 are taken at a point 0, and a plane cuts off intercepts proportional to $a_1/h_1, a_2/h_2, a_3/h_3$ on the axes. Show that this plane is perpendicular to the direction $h_1 b_1 + h_2 b_2 + h_3 b_3$.

[Camb. P.N.S. 1945]

24. Define the vector product of two vectors P and Q, and prove that

$$P \times (Q \times R) = Q(P.R) - R(P.Q).$$

Given three non-coplanar vectors P, Q, R, show how to write an arbitrary vector X in the form

$$X = (X.P')P + (X.Q')Q + (X.R')R.$$

Show that
$$(P' \times Q').R' = 1/(P \times Q).R,$$

and that
$$P = P', \quad Q = Q', \quad R = R',$$

if and only if P, Q, R form an orthogonal triad of unit vectors.

[Camb. M.T. 1960]

25. a_1, a_2, a_3 are three vectors. Vectors b_1, b_2, b_3 are defined as follows:

$$b_1 = \frac{a_2 \times a_3}{a_1.a_2 \times a_3}, \quad b_2 = \frac{a_3 \times a_1}{a_1.a_2 \times a_3}, \quad b_3 = \frac{a_1 \times a_2}{a_1.a_2 \times a_3}.$$

If
$$R = r_1 a_1 + r_2 a_2 + r_3 a_3 \quad \text{and} \quad K = k_1 b_1 + k_2 b_2 + k_3 b_3$$

show that
$$R.K = r_1 k_1 + r_2 k_2 + r_3 k_3.$$

If a_1, a_2, a_3 are unit vectors making angles of 60° with one another, find the angle between R and K in terms of $r_1, r_2, r_3, k_1, k_2, k_3$.

[Camb. N.S.I 1955]

26. Three vectors a_1, a_2, a_3 are given, and

$$b^1 = A^{-1} a_2 \times a_3, \quad b^2 = A^{-1} a_3 \times a_1, \quad b^3 = A^{-1} a_1 \times a_2,$$

where A denotes the triple product of $\mathbf{a}_1, \mathbf{a}_2, \mathbf{a}_3$. Prove that the triple product B of $\mathbf{b}^1, \mathbf{b}^2, \mathbf{b}^3$ is also A^{-1}. Prove also that

$$\mathbf{a}_1 = B^{-1}\mathbf{b}_2 \times \mathbf{b}^3, \quad \mathbf{a}_2 = B^{-1}\mathbf{b}^3 \times \mathbf{b}^1, \quad \mathbf{a}_3 = B^{-1}\mathbf{b}^1 \times \mathbf{b}^2,$$

and

$$\mathbf{a}_i.\mathbf{b}^k = \delta_i{}^k.$$

If $\mathbf{u}, \mathbf{v}$ are any two vectors, and $\mathbf{u} = \Sigma u^i \mathbf{a}_i = \Sigma u_i \mathbf{b}^i$, $\mathbf{v} = \Sigma v^i \mathbf{a}_i = \Sigma v_i \mathbf{b}^i$ where the summations run over the values $i = 1, 2, 3$ show that

$$\mathbf{u}.\mathbf{v} = \Sigma u^i v_i$$

$$\mathbf{u} \times \mathbf{v} = A(u^2 v^3 - u^3 v^2)\mathbf{b}^1 + \dots$$

[$\delta_i{}^k = 1$ when $i = k$ and zero otherwise.] [Camb. P.N.S. 1959]

27. $\mathbf{a}_1, \mathbf{a}_2, \mathbf{a}_3, \mathbf{a}_4$ are four vectors, $\mathbf{b}_1$ is the vector $\mathbf{a}_2 \times \mathbf{a}_3 + \mathbf{a}_3 \times \mathbf{a}_4 + \mathbf{a}_4 \times \mathbf{a}_2$, and $\mathbf{b}_2, \mathbf{b}_3, \mathbf{b}_4$ are obtained from $\mathbf{b}_1$ by a cyclic permutation of the suffixes 1, 2, 3, 4. Verify that each of the vectors

$$\mathbf{b}_1 - \mathbf{b}_2 + \mathbf{b}_3 - \mathbf{b}_4, \quad (\mathbf{a}_1 \times \mathbf{b}_1) - (\mathbf{a}_2 \times \mathbf{b}_2) + (\mathbf{a}_3 \times \mathbf{b}_3) - (\mathbf{a}_4 \times \mathbf{b}_4)$$

vanishes.

[Cf. Camb. M.T. II 1951]

28. Expand the following expressions:

$$(i)\ (\mathbf{l} \times \mathbf{m}).(\mathbf{l} \times \mathbf{n}) \qquad (ii)\ (\mathbf{l} \times \mathbf{m}) \times (\mathbf{l} \times \mathbf{n}).$$

Hence prove that the angle between two faces of a regular tetrahedron is $\cos^{-1} \frac{1}{3}$.

Prove that,

$$\epsilon_{ijk}\epsilon_{ilm} = \delta_{jl}\delta_{km} - \delta_{jm}\delta_{kl}$$

Hence or otherwise obtain an expansion for $\mathbf{A} \times (\mathbf{B} \times \mathbf{C})$.

[ϵ_{ijk} vanishes when two or more of i, j, k are the same, has value $+1$ when i, j, k is a cyclic permutation of 1, 2, 3 and -1 otherwise.]

[Camb. P.N.S. 1948]

Chapter 3

Vector Calculus

3.1. Vector function of a scalar

SUPPOSE that to each value of a scalar variable t there is assigned a vector $\mathbf{Q}$ in space. Then $\mathbf{Q}$ is said to be a vector function of the variable t. As t varies $\mathbf{Q}$ will generally vary both in magnitude and in direction. We may write $\mathbf{Q}$ as $\mathbf{Q}(t)$.

For example, if $\mathbf{r}$ is the position vector at time t of a moving particle, then $\mathbf{r}$ is a vector function of t.

When $\mathbf{Q}$ is expressed, in terms of its components along the axes of a fixed frame of reference, in the form

$$\mathbf{Q}(t) = Q_x(t)\mathbf{i} + Q_y(t)\mathbf{j} + Q_z(t)\mathbf{k} \tag{1}$$

where $\mathbf{i}, \mathbf{j}, \mathbf{k}$ are unit vectors along the three axes, the components Q_x, Q_y, Q_z are scalar functions of t.

For example,

$$\mathbf{Q} = t^2\mathbf{i} + \sin t\mathbf{j} + (1-t)\mathbf{k} \tag{2}$$

is a vector function of the scalar variable t.

The concept of the limit of a vector function as the variable t tends to t_0, and of the continuity of the function when t is equal to t_0, may be discussed in the same way as for scalar functions.

The differential coefficient or derivative of a vector function $\mathbf{Q}$ with respect to the scalar variable t is defined as the limit:

$$\frac{d\mathbf{Q}}{dt} = \lim_{\Delta t \to 0} \frac{\Delta \mathbf{Q}}{\Delta t} \tag{3}$$

where

$$\Delta \mathbf{Q} = \mathbf{Q}(t + \Delta t) - \mathbf{Q}(t). \tag{4}$$

VECTOR CALCULUS 39

It is useful to have a geometrical picture of the limiting process. Let O be the origin, and let the vector $\mathbf{Q}$ be represented by the line segment OP. As t varies continuously the extremity P of the segment OP will describe a curve (Fig. 1). Suppose at time t the extremity is at P in the figure and at time

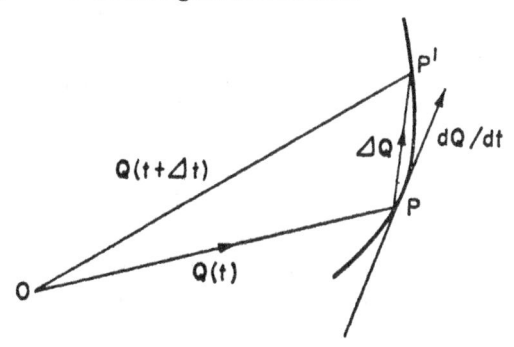

Fig. 1

$t + \Delta t$ the extremity is at P'. The vector $\Delta \mathbf{Q} = \mathbf{Q}(t + \Delta t) - \mathbf{Q}(t)$ is represented by the line segment PP'. $\Delta \mathbf{Q}/\Delta t$, being the product of the vector $\Delta \mathbf{Q}$ and the scalar $1/\Delta t$, is a vector. Therefore its limit as $\Delta t \to 0$, namely $d\mathbf{Q}/dt$, is also a vector. The direction of this vector will be along the limiting position of the chord PP' as P' tends to P, that is, along the tangent to the curve at P.

When $\mathbf{Q}$ is expressed in terms of its components, as in eqn. (1),

$$\Delta \mathbf{Q} = \mathbf{Q}(t + \Delta t) - \mathbf{Q}(t)$$
$$= Q_x(t + \Delta t)\mathbf{i} + Q_y(t + \Delta t)\mathbf{j} + Q_z(t + \Delta t)\mathbf{k}$$
$$- Q_x(t)\mathbf{i} - Q_y(t)\mathbf{j} - Q_z(t)\mathbf{k}$$
$$= \Delta Q_x \mathbf{i} + \Delta Q_y \mathbf{j} + \Delta Q_z \mathbf{k}.$$

Hence, proceeding to the limit we obtain

$$\frac{d\mathbf{Q}}{dt} = \frac{dQ_x}{dt}\mathbf{i} + \frac{dQ_y}{dt}\mathbf{j} + \frac{dQ_z}{dt}\mathbf{k}. \qquad (5)$$

We may write this equation as

$$\frac{d}{dt}(Q_x, Q_y, Q_z) = \left(\frac{dQ_x}{dt}, \frac{dQ_y}{dt}, \frac{dQ_z}{dt}\right). \quad (6)$$

The component, along any axis, of the derivative vector $d\mathbf{Q}/dt$ is equal to the derivative of the component of $\mathbf{Q}$ along that axis.

Second and higher derivatives may be defined as for scalar functions. Thus

$$\frac{d^2\mathbf{Q}}{dt^2} = \frac{d}{dt}\left(\frac{d\mathbf{Q}}{dt}\right). \quad (7)$$

If $\mathbf{r}$ is the position vector of a particle at time t, the derivative vector $d\mathbf{r}/dt$ is the velocity vector, and it is along the tangent to the path of the particle at the point corresponding to the instant t. The second derivative $d^2\mathbf{r}/dt^2$ is the acceleration vector at the instant t.

The following rules for the differentiation of vector functions, where $\mathbf{Q}$, $\mathbf{R}$ are vector functions of t and λ a scalar function of t, may be easily verified:

$$\frac{d}{dt}(\mathbf{Q} + \mathbf{R}) = \frac{d\mathbf{Q}}{dt} + \frac{d\mathbf{R}}{dt} \quad (8)$$

$$\frac{d}{dt}(\lambda\mathbf{Q}) = \lambda\frac{d\mathbf{Q}}{dt} + \frac{d\lambda}{dt}\mathbf{Q} \quad (9)$$

$$\frac{d}{dt}(\mathbf{Q}\cdot\mathbf{R}) = \mathbf{Q}\cdot\frac{d\mathbf{R}}{dt} + \frac{d\mathbf{Q}}{dt}\cdot\mathbf{R} \quad (10)$$

$$\frac{d}{dt}(\mathbf{Q}\times\mathbf{R}) = \mathbf{Q}\times\frac{d\mathbf{R}}{dt} + \frac{d\mathbf{Q}}{dt}\times\mathbf{R}. \quad (11)$$

For example, to prove the relation (10), suppose $\mathbf{Q} + \Delta\mathbf{Q}$ and $\mathbf{R} + \Delta\mathbf{R}$ are the values of $\mathbf{Q}$ and $\mathbf{R}$ at time $t + \Delta t$. Then

$$\frac{(\mathbf{Q} + \Delta\mathbf{Q}).(\mathbf{R} + \Delta\mathbf{R}) - \mathbf{Q}.\mathbf{R}}{\Delta t} \tag{12}$$

$$= \mathbf{Q}.\frac{\Delta\mathbf{R}}{\Delta t} + \frac{\Delta\mathbf{Q}}{\Delta t}.\mathbf{R} + \frac{\Delta\mathbf{Q}.\Delta\mathbf{R}}{\Delta t}, \tag{13}$$

which, in the limit as Δt tends to zero, tends to

$$\mathbf{Q}.\frac{d\mathbf{R}}{dt} + \frac{d\mathbf{Q}}{dt}.\mathbf{R}, \tag{14}$$

the last term in (13) vanishing in the limit.

In general, the differentiation of vectors follows the same rules as in ordinary differential calculus, except that the order of factors in a vector product should be carefully maintained, as for example in the eqn. (11) when $\mathbf{Q}$ is kept to the left of $\mathbf{R}$ in every term of the equation.

We note that in general

$$\left|\frac{d\mathbf{Q}}{dt}\right| \neq \frac{d}{dt}|\mathbf{Q}|, \tag{15}$$

that is, the magnitude of the derivative vector is not in general equal to the derivative of the magnitude of the vector.

From the relation

$$\mathbf{Q}.\mathbf{Q} = \mathbf{Q}^2 = |\mathbf{Q}|^2 \tag{16}$$

we obtain by differentiation

$$2\mathbf{Q}.\frac{d\mathbf{Q}}{dt} = 2|\mathbf{Q}|\frac{d}{dt}|\mathbf{Q}|,$$

that is,

$$\frac{d}{dt}|\mathbf{Q}| = \frac{\mathbf{Q}}{|\mathbf{Q}|}.\frac{d\mathbf{Q}}{dt} \tag{17}$$

$\mathbf{Q}/|\mathbf{Q}|$ is a unit vector in the direction of $\mathbf{Q}$, and we may interpret the eqn. (17) as implying that $(d/dt)|\mathbf{Q}|$ is equal to the component of $d\mathbf{Q}/dt$ in the direction of $\mathbf{Q}$.

3.2. Unit tangent vector T

An example of the derivative of a vector function with respect to a scalar variable is provided by the unit tangent vector to a curve in three-dimensional space.

Suppose that P_0 is a fixed point on a curve, and P a variable point on the curve with position vector **r** (Fig. 2).

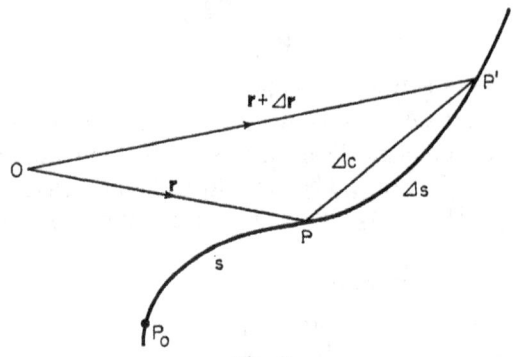

Fig. 2

Let the arc length P_0P be denoted by s. Let P' be a neighbouring point to P on the curve and have position vector $\mathbf{r} + \Delta\mathbf{r}$. Let Δs be the arc distance PP' and Δc the length of the chord PP'. The segment $\overrightarrow{PP'}$, denotes the vector $\Delta\mathbf{r}$ whose length is Δc, and therefore $(\Delta c)^2 = (\Delta\mathbf{r})^2$. Since $\Delta\mathbf{r}$ is a vector along PP' and $1/\Delta s$ is a scalar quantity $\Delta\mathbf{r}/\Delta s$ is a vector in the direction of PP'. As $\Delta s \to 0$, the point P' tends to the point P, and the chord PP' tends to the tangent to the curve at P. Also, $\Delta c/\Delta s$ tends to the limit 1 as $\Delta s \to 0$. Hence

$$\left(\frac{\Delta\mathbf{r}}{\Delta s}\right)^2 = \left(\frac{\Delta c}{\Delta s}\right)^2 \to 1 \text{ as } \Delta s \to 0.$$

Therefore, the vector

$$\frac{d\mathbf{r}}{ds} = \lim_{\Delta s \to 0} \frac{\Delta\mathbf{r}}{\Delta s}$$

has unit magnitude, and is along the tangent to the curve at P. Hence

$$\mathbf{T} = \frac{d\mathbf{r}}{ds} = \left(\frac{dx}{ds}, \frac{dy}{ds}, \frac{dz}{ds}\right) \tag{18}$$

is unit vector along the tangent to the curve at P.

Example. The equations of a curve in three-dimensional space are given in terms of a parameter θ by

$$x = a\cos\theta, \quad y = a\sin\theta, \quad z = c\theta$$

where (x, y, z) are rectangular coordinates and a, c are constants. Find the unit tangent vector at the point corresponding to the parameter θ.

If s denotes the arc distance from a fixed point on the curve, the equation

$$\left|\frac{d\mathbf{r}}{ds}\right|^2 = 1$$

gives

$$ds^2 = dx^2 + dy^2 + dz^2 \tag{19}$$
$$= a^2 \sin^2\theta (d\theta)^2 + a^2 \cos^2\theta (d\theta)^2 + c^2 (d\theta)^2$$
$$= (a^2 + c^2)(d\theta)^2.$$

Hence

$$\frac{d\theta}{ds} = \frac{1}{\sqrt{(a^2 + c^2)}} \tag{20}$$

taking the positive value of the root, by measuring s in the sense of increasing θ. The unit tangent vector at the point whose parameter is θ is

$$\mathbf{T} = \left(\frac{dx}{ds}, \frac{dy}{ds}, \frac{dz}{ds}\right) = \left(-a\sin\theta \frac{d\theta}{ds}, a\cos\theta \frac{d\theta}{ds}, c\frac{d\theta}{ds}\right)$$
$$= \left[-\frac{a}{\sqrt{(a^2 + c^2)}}\sin\theta, \frac{a}{\sqrt{(a^2 + c^2)}}\cos\theta, \frac{c}{\sqrt{(a^2 + c^2)}}\right]. \tag{21}$$

3.3. Functions of a vector

Suppose that ϕ is a scalar quantity which is known whenever a vector variable is given. Denoting the vector variable by **r**, the function ϕ is said to be a single valued scalar function of the vector **r** and may be written

$$\phi(\mathbf{r}). \tag{22}$$

We may regard **r** as the position vector, of a variable point P in a space of three dimensions with respect to a fixed origin O so that $\overrightarrow{OP} = \mathbf{r}$. ϕ is then represented as a function of the variable point P and is denoted by the symbol $\phi(P)$ or by $\phi(x, y, z)$ where x, y, z are the components of **r** in a rectangular coordinate system.

The function $\phi(P)$, which varies as the point P varies, is said to define a *field*, and because in this case the function ϕ is a scalar, this field is said to be a *scalar field*. For example, $\phi(P)$ could be the temperature at different points P of a gas. Temperature is a scalar quantity, and the temperature field is an example of a scalar field.

In the same way we may define single valued vector functions of position:

$$\mathbf{Q}(P) \quad \text{or} \quad \mathbf{Q}(\mathbf{r}). \tag{23}$$

The field which such a function defines is a *vector field*. Such a field associates with each point of space a vector quantity which has a magnitude as well as a direction. Such a vector is sometimes also known as a point vector, being a function of the point P.

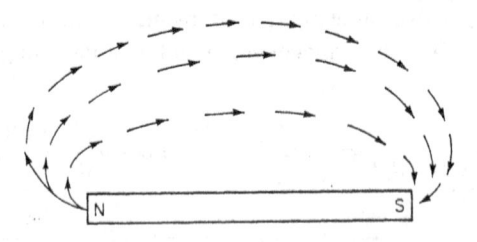

Fig. 3

VECTOR CALCULUS

An instance of a vector field is provided by a magnetic field. An experiment with a bar magnet and iron filings will show the association with each point a magnetic field **Q** which has both direction and magnitude. At different points in space, the magnetic field has different magnitudes and directions (Fig. 3).

Other familiar examples of a vector field are the electrostatic force field, the gravitational force field, and the velocity field of a moving fluid.

3.4. Map of a field

We may map a scalar field $\phi(\mathbf{r})$ in three-dimensional space by drawing the family of surfaces

$$\phi(\mathbf{r}) = \text{constant} \tag{24}$$

for different values of the constant.

Two such surfaces will in general not intersect. Suppose that this property does not hold and that two surfaces $\phi = C_1$ and $\phi = C_2$ intersect, where C_1 and C_2 are constants and P is a point of intersection. Since P lies on the first surface $\phi(P) = C_1$, and since P lies on the second surface $\phi(P) = C_2$. These conditions cannot both be simultaneously satisfied by any point P if ϕ is a single valued function unless $C_1 = C_2$.

These surfaces are known as *level surfaces*. We may divide the region of space under consideration by a system of non-intersecting level surfaces (Fig. 4).

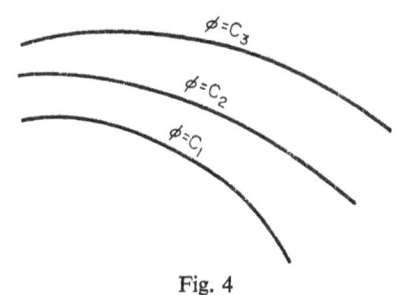

Fig. 4

Illustration 1. The potential energy of a particle of mass m, in the gravitational field of the earth near the earth's surface, is of the form

$$\phi = mg(z - z_0) \tag{25}$$

where the z-axis is vertically upwards, g is the acceleration due to gravity and z_0 is a constant. The level surfaces are given by

$$mg(z - z_0) = \text{constant},$$

that is,

$$z = \text{constant.} \tag{26}$$

The level surfaces here are a system of horizontal planes (Fig. 5) where k_1 and k_2 are positive constants.

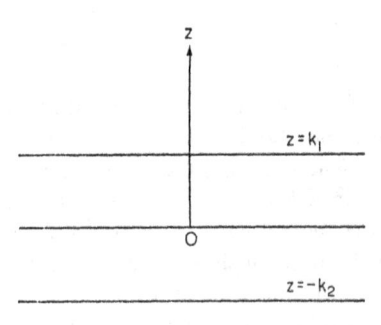

Fig. 5

Illustration 2. The electrostatic potential at a point P due to an electric charge e at A and another charge e at B is

$$\phi = \frac{e}{r} + \frac{e}{r'} \tag{27}$$

where $r = AP$ and $r' = BP$. The level surfaces here are called equipotential surfaces and are given by

$$\frac{1}{r} + \frac{1}{r'} = \text{constant.} \tag{28}$$

Let us consider a section of these surfaces by a plane through
AB. We obtain a system of curves of the sixth degree. A rough
sketch of these curves is given in Fig. 6.

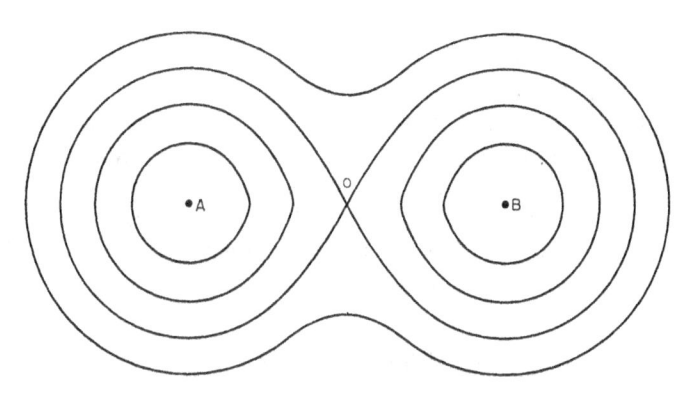

Fig. 6

The equipotential which passes through the point O, which is the
mid-point of AB, has some special features. O is an exceptional
point in that the equipotential through it intersects itself at O.
At O,

$$\frac{\partial \phi}{\partial x} = \frac{\partial \phi}{\partial y} = \frac{\partial \phi}{\partial z} = 0.$$

It is a point of equilibrium. Near O the equipotential surface is a
double tangent cone. For large positive values of the constant
ϕ the equipotential curves are small closed curves surrounding
A or B. When $\phi = 2e/a$, we obtain the special curve through
O. When ϕ is smaller than $2e/a$, the equipotential curve is a large
closed curve surrounding A, B and O. As ϕ becomes smaller, the
corresponding curve becomes larger and larger.

It is also of use to map vector fields. Given a vector field $\mathbf{A}$,
an $\mathbf{A}$-line is defined as a curve whose tangent at any pointon it

is the direction of the vector **A** at that point. An **A**-line satisfies the differential equations

$$\frac{\mathrm{d}x}{A_x} = \frac{\mathrm{d}y}{A_y} = \frac{\mathrm{d}z}{A_z}.$$

The **A**-line drawn through the points of any closed curve C will form a tubular surface called an **A**-tube.

A familiar example of **A**-lines and **A**-tubes is obtained when for the vector **A** we take the magnetostatic field vector **H**. The **H**-lines are usually called lines of force, and the **H**-tubes are tubes of force. If we take a bar magnet and sprinkle iron filings near it, the lines of force may be seen clearly, as originating from one pole of the magnet and ending on the other pole (see Fig. 3).

3.5. Directional derivative

Given a function

$$\phi(P) = \phi(\mathbf{r}) = \phi(x, y, z)$$

we may define the partial derivatives

$$\frac{\partial \phi}{\partial x}, \frac{\partial \phi}{\partial y}, \frac{\partial \phi}{\partial z}$$

where

$$\frac{\partial \phi}{\partial x} = \lim_{\Delta x \to 0} \frac{\phi(x + \Delta x, y, z) - \phi(x, y, z)}{\Delta x} \tag{29}$$

and is the derivative with respect to x where y and z remain constant, that is, $\partial \phi / \partial x$ is the rate of change along a line parallel to the axis of x. Similarly $\partial \phi / \partial y$ and $\partial \phi / \partial z$ are the rates of change along parallels to the axes of y and z respectively.

We may define the directional derivative along an arbitrary direction as

$$\lim_{\Delta s \to 0} \frac{\Delta \phi}{\Delta s} \tag{30}$$

VECTOR CALCULUS 49

where $\Delta\phi$ is the change in ϕ when the point P is moved a distance Δs in the particular direction.

We establish the result that the directional derivative of ϕ at the point $P_0(x_0, y_0, z_0)$ along the line l which has direction cosines $\cos\alpha$, $\cos\beta$, $\cos\gamma$ is

$$\left(\frac{\partial\phi}{\partial x}\right)_0 \cos\alpha + \left(\frac{\partial\phi}{\partial y}\right)_0 \cos\beta + \left(\frac{\partial\phi}{\partial z}\right)_0 \cos\gamma. \tag{31}$$

The coordinates of a point on the line l at a distance s from P_0 are

$$x = x_0 + s\cos\alpha, \quad y = y_0 + s\cos\beta, \quad z = z_0 + s\cos\gamma. \tag{32}$$

At points on the line l, ϕ is a function of s.

Hence

$$\lim_{\Delta s \to 0} \frac{\Delta\phi}{\Delta s}.$$

is the usual differential coefficient of ϕ with respect to s, namely $(\mathrm{d}\phi/\mathrm{d}s)_0$.

Thus the directional derivative of ϕ at P_0 along l is $(\mathrm{d}\phi/\mathrm{d}s)_0$. By the standard rule of partial differentiation

$$\frac{\mathrm{d}\phi}{\mathrm{d}s} = \frac{\partial\phi}{\partial x}\frac{\mathrm{d}x}{\mathrm{d}s} + \frac{\partial\phi}{\partial y}\frac{\mathrm{d}y}{\mathrm{d}s} + \frac{\partial\phi}{\partial z}\frac{\mathrm{d}z}{\mathrm{d}s}. \tag{33}$$

From (32),

$$\frac{\mathrm{d}x}{\mathrm{d}s} = \cos\alpha, \qquad \frac{\mathrm{d}y}{\mathrm{d}s} = \cos\beta, \qquad \frac{\mathrm{d}z}{\mathrm{d}s} = \cos\gamma. \tag{34}$$

Hence the directional derivative is

$$\left(\frac{\mathrm{d}\phi}{\mathrm{d}s}\right)_0 = \left(\frac{\partial\phi}{\partial x}\right)_0 \cos\alpha + \left(\frac{\partial\phi}{\partial y}\right)_0 \cos\beta + \left(\frac{\partial\phi}{\partial z}\right)_0 \cos\gamma. \tag{35}$$

One may also define a "directional derivative along a curve". This is the derivative at a point on a given curve along the tangent to the curve at that point. If s is the arc distance of a point P

on the curve from a fixed point on the curve, the unit vector along the tangent is

$$\mathbf{T} = \frac{d\mathbf{r}}{ds} = \left(\frac{dx}{ds}, \frac{dy}{ds}, \frac{dz}{ds}\right).$$

The coordinates of P being functions of s, $\phi(P)$ for points on the curve is a function of s. The directional derivative of ϕ at P along the curve is, as in the case of the straight line l considered earlier, $d\phi/ds$, which by rules of partial differentiation is

$$\frac{\partial \phi}{\partial x}\frac{dx}{ds} + \frac{\partial \phi}{\partial y}\frac{dy}{ds} + \frac{\partial \phi}{\partial z}\frac{dz}{ds} = \frac{\partial \phi}{\partial x}T_x + \frac{\partial \phi}{\partial y}T_y + \frac{\partial \phi}{\partial z}T_z. \quad (36)$$

3.6. Gradient vector

An important property of the partial derivatives $\partial\phi/\partial x$, $\partial\phi/\partial y$, $\partial\phi/\partial z$ of a scalar function ϕ is that they are the resolutes, along the axes of coordinates, of a vector which has many important uses.

We may verify that the directed quantity whose components are $\partial\phi/\partial x$, $\partial\phi/\partial y$, $\partial\phi/\partial z$ satisfies the criteria for a vector quantity. Apart from having magnitude and direction, the quantity must obey the vector law of addition. If ϕ_1 and ϕ_2 are scalar functions,

$$\frac{\partial}{\partial x}(\phi_1 + \phi_2) = \frac{\partial \phi_1}{\partial x} + \frac{\partial \phi_2}{\partial x},$$

with similar relations for the derivatives with respect to y and z. From this property we may establish that such quantities obey the rule for vector addition.

If we denote the vector whose components are $\partial\phi/\partial x$, $\partial\phi/\partial y$, $\partial\phi/\partial z$ by $\mathbf{G}$, then

$$G_x = \frac{\partial \phi}{\partial x}, \qquad G_y = \frac{\partial \phi}{\partial y}, \qquad G_z = \frac{\partial \phi}{\partial z}. \quad (37)$$

VECTOR CALCULUS

The directional derivative of a scalar function ϕ along a curve (or straight line) may be expressed in terms of the vector **G** in the following way. From the eqn. (36), the derivative is

$$\frac{d\phi}{ds} = \mathbf{G}.\mathbf{T}, \tag{38}$$

where **T** is unit vector along the tangent to the curve at the point under consideration. (In the case of a straight line, **T** is unit vector along the line).

We now state and prove a theorem concerning the vector **G**.

*The vector **G** at any point is normal to the level surface $\phi = $ constant which passes through that point.*

Suppose S is the surface whose equation is $\phi = C$ and which passes through the point P. (Fig. 7). Consider a curve which

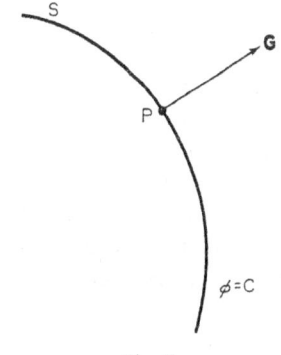

Fig. 7

passes through P and is such that the tangent at P to the curve lies on the tangent plane at P to the surface S. Along such a curve, the directional derivative at P is $d\phi/ds$ which vanishes since the equation of the surface is $\phi = C$. If **T** denotes the unit vector along the tangent to the curve at P, then eqn. (38) gives

$$\mathbf{G}.\mathbf{T} = 0. \tag{39}$$

Hence **G** is perpendicular to **T**. This property holds for every curve whose tangent at P lies on the tangent plane at P to the surface S.

This vector **G** is called the gradient of ϕ and is written as grad ϕ.

$$\mathbf{G} = \text{grad } \phi = \left(\frac{\partial \phi}{\partial x}, \frac{\partial \phi}{\partial y}, \frac{\partial \phi}{\partial z}\right). \tag{40}$$

Also

$$|\mathbf{G}| = \sqrt{\left\{\left(\frac{\partial \phi}{\partial x}\right)^2 + \left(\frac{\partial \phi}{\partial y}\right)^2 + \left(\frac{\partial \phi}{\partial z}\right)^2\right\}}. \tag{41}$$

The directional derivative of ϕ along any line l may be expressed in terms of the gradient vector in the following way:

From (31), the derivative along the line l, the direction cosines of which are $(\cos \alpha, \cos \beta, \cos \gamma)$, is

$$\frac{d\phi}{ds} = \mathbf{G}_x \cos \alpha + \mathbf{G}_y \cos \beta + \mathbf{G}_z \cos \gamma$$

$$= \text{comp}_l \mathbf{G}. \tag{42}$$

Hence we obtain the theorem:

The component of grad ϕ in any direction is equal to the directional derivative of ϕ in that direction.

Though initially we defined **G** in relation to a particular set of axes of coordinates, we see that it is a quantity that is independent of the choice of axes. Given the scalar field ϕ, we obtain a vector field **G** which is known at every point in the region under consideration.

The following illustration shows the important geometrical property of the gradient vector **G**. Consider two neighbouring surfaces S and S' when S has equation $\phi = C$ and S' has equation $\phi = C + \delta C$. Suppose the normal at P to the surface S meets the surface S' at Q (Fig. 8). Let a given curve through P meet the surface S' at P', and let l the tangent to the curve at P make an angle θ with the normal PQ.

$$|\mathbf{G}| = \lim_{\delta C \to 0} \frac{\delta C}{PQ}. \tag{43}$$

VECTOR CALCULUS 53

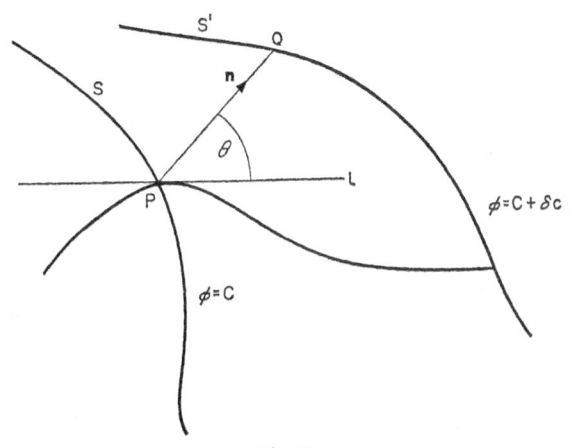

Fig. 8

The vector **G** is normal at P to S, and its magnitude is the rate of change of ϕ with respect to distance, in the direction of the normal at P. If **n** denotes the normal to S at P, it is a convenient notation to denote the derivative along the normal by $\partial \phi/\partial \mathbf{n}$. We then have

$$\mathbf{G} = \frac{\partial \phi}{\partial \mathbf{n}}, \qquad |\mathbf{G}| = \frac{\partial \phi}{\partial n}. \tag{44}$$

The directional derivative along the curve PP' is the limit of $\delta C/PP'$ as $\delta C \to 0$.

$$\frac{\delta C}{PP'} = \frac{\delta C}{PQ} \cdot \frac{PQ}{PP'}. \tag{45}$$

Proceeding to the limit, we obtain

$$\frac{\mathrm{d}\phi}{\mathrm{d}s} = |\mathbf{G}| \cos \theta,$$

that is, the directional derivative along l is the $\text{comp}_l \mathbf{G}$.

The gradient vector grad ϕ has its direction along that of the greatest increase of ϕ and its magnitude is the value of this maximum increase.

Example 1. Evaluate the directional derivative of the function

$$\phi = x^2 - y^2 + 2z^2$$

at the point P (1, 2, 3) in the direction of the line PQ where Q has coordinates (5, 0, 4).

$$\frac{\partial \phi}{\partial x} = 2x, \qquad \frac{\partial \phi}{\partial y} = -2y, \qquad \frac{\partial \phi}{\partial z} = 4z.$$

Hence at P, the components of **G** are (2, −4, 12). Direction cosines of PQ are proportional to 5 − 1, 0 − 2, 4 − 3, that is, to 4, −2, 1. Hence the direction cosines of PQ are

$$\frac{4}{\sqrt{(4^2 + 2^2 + 1^2)}}, \frac{-2}{\sqrt{(4^2 + 2^2 + 1^2)}}, \frac{1}{\sqrt{(4^2 + 2^2 + 1^2)}},$$

that is

$$\frac{4}{\sqrt{21}}, \frac{-2}{\sqrt{21}}, \frac{1}{\sqrt{21}}.$$

The required directional derivative is

$$\mathbf{G}_x \cos \alpha + \mathbf{G}_y \cos \beta + \mathbf{G}_z \cos \gamma$$

$$= 2\left(\frac{4}{\sqrt{21}}\right) + (-4)\left(\frac{-2}{\sqrt{21}}\right) + 12\left(\frac{1}{\sqrt{21}}\right)$$

$$= \tfrac{4}{3}\sqrt{21}.$$

Example 2. Find the components of grad$(1/r)$ where $r = (x^2 + y^2 + z^2)^{\frac{1}{2}}$. $\partial/\partial x(1/r) = -(1/r^2)\partial r/\partial x$. From $r^2 = x^2 + y^2 + z^2$, differentiating partially with respect to x, we obtain

$$2r \frac{\partial r}{\partial x} = 2x, \qquad \text{hence} \quad \frac{\partial r}{\partial x} = \frac{x}{r}.$$

Substituting, we have

$$\frac{\partial}{\partial x}\left(\frac{1}{r}\right) = -\frac{x}{r^3}.$$

We obtain similar expressions for $\dfrac{\partial}{\partial y}\left(\dfrac{1}{r}\right)$ and $\dfrac{\partial}{\partial z}\left(\dfrac{1}{r}\right)$.

Thus
$$\operatorname{grad}\left(\frac{1}{r}\right) = \left[\frac{-x}{r^3}, \frac{-y}{r^3}, \frac{-z}{r^3}\right].$$

Exercise III

1. If **r** is a vector function of a scalar t and **a** a constant vector, differentiate with respect to t

   (i) **r.a**   (ii) **r** × **a**   (iii) **r.ṙ**   (iv) **r** × **ṙ**

   (v) **r.a**   (vi) [**a, r, ṙ**]   (vii) [**r, ṙ, r̈**]

2. If **A, B** are constant vectors, and **r** a vector function of the scalar variable t given by

$$\mathbf{r} = \mathbf{A}\cos\lambda t + \mathbf{B}\sin\lambda t$$

where λ is a constant, show that

$$\frac{d^2\mathbf{r}}{dt^2} + \lambda^2\mathbf{r} = \mathbf{O}.$$

3. Find unit vector **T** for the curve

$$\mathbf{r} = a\cos t\,\mathbf{i} + a\sin t\,\mathbf{j} + ct\,\mathbf{k}.$$

4. A particle moves with constant velocity **v**, starting from a point with position vector $\mathbf{r}_0$ at time $t = 0$. Give the position vector **r** of the particle at time t and show that the particle will be (or was) closest to the coordinate origin at time

$$t = -\frac{\mathbf{r}_0 \cdot \mathbf{v}}{\mathbf{v} \cdot \mathbf{v}}.$$

[Camb. N.S. 1954]

5. If **l, m** are mutually perpendicular unit vectors which rotate about the instantaneous direction of their common normal with angular velocity ω (reckoned positive right-handedly about **l** × **m**), show that

$$\frac{d\mathbf{l}}{dt} = \omega\mathbf{m}, \qquad \frac{d\mathbf{m}}{dt} = -\omega\mathbf{l}.$$

Hence show that the direction of the common normal to **l** and **m** remains fixed.

[Camb. P.N.S. 1954]

6. The vector **r** satisfies the vector equation

$$m\frac{d^2\mathbf{r}}{dt^2} = e\mathbf{E} + \frac{e}{c}\frac{d\mathbf{r}}{dt} \times \mathbf{H}$$

where $\mathbf{E} = (0, E, 0)$, $\mathbf{H} = (0, 0, H)$ and $e, m, c, \mathbf{E}, \mathbf{H}$ are constants. Write the equation in component form and show by solving the equation that

$$x = \frac{cEt}{H} - \frac{mc^2E}{eH^2}\sin\left(\frac{eHt}{mc}\right), \qquad y = \frac{mc^2E}{eH^2}\left\{1 - \cos\left(\frac{eHt}{mc}\right)\right\}, \qquad z = 0$$

is the solution that satisfies the conditions

$$\mathbf{r} = \mathbf{O}, \qquad \frac{d\mathbf{r}}{dt} = \mathbf{O} \quad \text{at} \quad t = 0.$$

[Camb. P.N.S. 1947]

7. Sketch the equipotential surfaces $\phi = $ const., where

$$O = (0, 0, 0), \quad A = (-a, 0, 0), \quad B = (a, 0, 0), \quad P = (x, y, z).$$

and

(i) $\phi = \dfrac{1}{OP}$ (ii) $\phi = \dfrac{1}{AP} - \dfrac{1}{BP}$ (iii) $\phi = \dfrac{1}{AP} + \dfrac{1}{BP} - \dfrac{2}{OP}$

(iv) $\phi = \log\left|\dfrac{AP + BP + 2a}{AP + BP - 2a}\right|.$

8. If $\mathbf{r}_1$, $\mathbf{r}_2$ are the vectors from points A_1 and A_2 to a point P, and if $r_1 = |\mathbf{r}_1|, r_2 = |\mathbf{r}_2|$ show that for the ellipse $r_1 + r_2 = $ constant,

$$\mathbf{T} = \frac{d\mathbf{r}_1}{ds} = \frac{dr_1}{ds}\mathbf{u} + r_1\frac{d\mathbf{u}}{ds} \text{ where } \mathbf{u} \text{ is unit vector along } \mathbf{r}_1.$$

Hence show that

$$\mathbf{T}\cdot\mathbf{u} = \frac{dr_1}{ds}$$

and that the tangent vector makes equal angles with A_1P and A_2P.

9. $A = -(c, 0)$, $B = (c, 0)$, $\phi = \log AP/BP$.

If $\mathbf{G} = \text{grad } \phi$, find the field lines given by $dx/G_x = dy/G_y$, and show that they intersect the surfaces $\phi = $ const. orthogonally.

10. Find the directional derivative at B of

$$\phi = x^4 + y^4 + z^4$$

along AB where $B = (2, 6, -1)$, $A = (1, -2, 1)$.

11. Given the surfaces

$$x^2 + y^2 + z^2 = 36, \quad x^2 - y^2 + z^2 = 16$$

find the directional derivative of

$$\phi = x^2 - 3y^2 + z^2$$

at the point $(4, \sqrt{10}, \sqrt{10})$ in the direction of the curve of intersection of the surfaces.

12. Find the directional derivative of

$$\phi = (x^2 + y^2 + z^2)^{-1/2}$$

at the point $(3, 1, 2)$ in the direction of the vector

$$yz\mathbf{i} + zx\mathbf{j} + xy\mathbf{k}.$$

Chapter 4

Vector Calculus

4.1. Line integrals

SUPPOSE C is an arc of a curve with terminal points A and B (Fig. 1). Let $\phi(P)$ be a scalar point function. The integral of ϕ over the curve C may be defined in the following ways. The arc is divided

Fig. 1

into m parts by $m+1$ points $P_0, P_1, \ldots, P_m$ where $A \equiv P_0$, $B \equiv P_m$. Let Δs_q where $q = 1, 2, \ldots, m$ be the length of the arc $P_{q-1} P_q$, and let X_q be point on the curve within the arc $P_{q-1} P_q$. The expression

$$\phi(X_q)\Delta s_q \qquad (1)$$

is formed for each interval Δs_q, and the different expressions as q takes the values $1, 2, \ldots, m$ are summed, thus yielding

$$\phi(X_1)\Delta s_1 + \phi(X_2)\Delta s_2 + \ldots + \phi(X_m)\Delta s_m. \qquad (2)$$

If as $m \to \infty$, and each $\Delta s_q \to 0$, the expression (2) tends to a finite limit which is independent of the mode of division of C into its elements Δs_q, the value of this limit is called the integral of ϕ over C and is written

$$\int_C \phi \, ds. \tag{3}$$

Such an integral over a curve C is called a *line integral* or a *curvilinear integral* and the curve C is called the *path of integration*.

It will be noted that this is a simple generalization of the Riemann integral of a function of a single variable

$$\int_a^b f(x) \, dx, \tag{4}$$

where $f(x)$ is a function of the variable x, and the limits of integration are a and b, that is, the path of integration is along the x-axis from the point $x = a$ to the point $x = b$.

The existence of the limit of the sum in expression (2) may be shown, in respect of the functions which we usually use in Applied Mathematics, by the methods used in respect of the integral (4) and will be assumed in the rest of this book. The functions we deal with are assumed to be single valued, continuous and differentiable. In practice the evaluation of the integral would not be through the direct use of the limiting process. Like integral (4), it is determined by utilizing its connection with the indefinite integral.

The integrands in line integrals may be scalars or vectors. Thus there are integrals of the type

$$\int \mathbf{A} \, ds \tag{5}$$

which in terms of Cartesian components may be written

$$\mathbf{i} \int A_x \, ds + \mathbf{j} \int A_y \, ds + \mathbf{k} \int A_z \, ds. \tag{6}$$

If s denotes the arc distance P_0P of a variable point P on the curve from a fixed point P_0 on the curve, the symbol d**s** may be used to denote the vector whose magnitude is the infinitesimal scalar element ds and whose direction is along the tangent to the curve at P. That is,

$$d\mathbf{s} = \mathbf{T}\,ds = \mathbf{i}\,dx + \mathbf{j}\,dy + \mathbf{k}\,dz \tag{7}$$

where

$$\mathbf{T} = \frac{d\mathbf{r}}{ds} \tag{8}$$

is unit vector along the tangent at P.

The scalar line integral of a vector $\mathbf{A}$ over a curve C is defined as

$$\int_C \mathbf{A}\cdot d\mathbf{s}. \tag{9}$$

We note that

$$\int_C \mathbf{A}\cdot d\mathbf{s} = \int_C A_s\,ds \tag{10}$$

where A_s is the component of $\mathbf{A}$ in the direction of the tangent to the curve at P (Fig. 2). In terms of Cartesian components (9) may be written

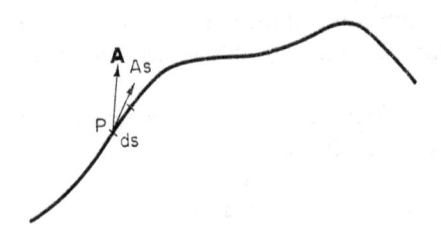

Fig. 2

$$\int_C (A_x\mathbf{i} + A_y\mathbf{j} + A_z\mathbf{k})\cdot(\mathbf{i}\,dx + \mathbf{j}\,dy + \mathbf{k}\,dx)$$
$$= \int_C (A_x\,dx + A_y\,dy + A_z\,dz). \tag{11}$$

A vector line integral may be defined in the same way, namely

$$\int_C \mathbf{A} \times d\mathbf{s} \tag{12}$$

which in terms of Cartesian components may be written

$$\mathbf{i}\int_C (A_y\,dz - A_z\,dy) + \mathbf{j}\int_C (A_z\,dx - A_x\,dz) + \mathbf{k}\int_C (A_x\,dy - A_y\,dx). \tag{13}$$

Example 1. Evaluate

$$\int_C (x + y + z)\,ds$$

where C the path of integration is

(a) the straight line joining $A(0, 0, 0)$ to $B(6, 8, 10)$;
(b) the arc of the circle $x^2 + y^2 = a^2$, $z = 0$ in the first quadrant in the xy-plane.

(a) The equation of the straight line AB is $x/3 = y/4 = z/5$.

The coordinates of any point on the line may be expressed in terms of a parameter t by the equations

$$x = 3t, \qquad y = 4t, \qquad z = 5t$$

where t takes values from $t = 0$ to $t = 2$.

The differentials dx, dy, dz are expressed in terms of dt by

$$dx = 3\,dt, \qquad dy = 4\,dt, \qquad dz = 5\,dt.$$

Hence

$$d\mathbf{s} = \mathbf{i}\,dx + \mathbf{j}\,dy + \mathbf{k}\,dz$$

$$= (3\mathbf{i} + 4\mathbf{j} + 5\mathbf{k})\,dt.$$

Therefore

$$ds = |d\mathbf{s}| = \sqrt{(3^2 + 4^2 + 5^2)}\,dt = 5\sqrt{2}\,dt.$$

$$\int_C \phi\,ds = \int_C (x + y + z)\,ds = \int_0^2 (3t + 4t + 5t)5\sqrt{2}\,dt$$

$$= 60\sqrt{2} \int_0^2 t\,dt = 120\sqrt{2}.$$

(b) The coordinates of any point on the arc of the circle $x^2 + y^2 = a^2$, $z = 0$ in the first quadrant may be expressed in terms of a parameter θ where

$$x = a \cos \theta, \qquad y = a \sin \theta, \qquad z = 0$$

where $0 \leqslant \theta \leqslant \pi/2$.

$$\mathrm{d}\mathbf{s} = \mathbf{i}\,\mathrm{d}x + \mathbf{j}\,\mathrm{d}y + \mathbf{k}\,\mathrm{d}z$$
$$= \mathbf{i}(-a \sin \theta\, \mathrm{d}\theta) + \mathbf{j}(a \cos \theta\, \mathrm{d}\theta)$$
$$\mathrm{d}s^2 = (-a \sin \theta\, \mathrm{d}\theta)^2 + (a \cos \theta\, \mathrm{d}\theta)^2 = (a\, \mathrm{d}\theta)^2.$$

Hence $\mathrm{d}s = a\, \mathrm{d}\theta$.

$$\int_C \phi\, \mathrm{d}s = \int_C (x + y + z)\, \mathrm{d}s = \int_0^{\pi/2} (a \cos \theta + a \sin \theta) a\, \mathrm{d}\theta$$
$$= a^2 [\sin \theta - \cos \theta]_0^{\pi/2} = 2a^2.$$

Example 2. A smooth wire in the form of the cycloid

$$s = 4a \sin \psi, \qquad (0 \leqslant \psi \leqslant \pi/2)$$

is fixed with its axis vertical and vertex 0 at the highest point. A bead of mass m free to slide along the wire under gravity, is released from rest at the point A where $\psi = \pi/6$. Find the time that the bead takes to reach the lowest point B where $\psi = \pi/2$.

Let v be the speed at a point P of the cycloid where $OP = s$. Then $\mathrm{d}s/\mathrm{d}t = v$, and hence

$$\mathrm{d}t = \frac{1}{v}\, \mathrm{d}s.$$

The required time is

$$T = \int_C \frac{\mathrm{d}s}{v}.$$

This is a line integral where C, the path of integration, is the part of the cycloid from A to B. To evaluate the integral we need to

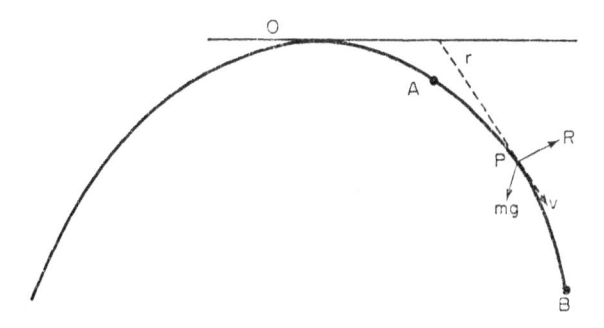

Fig. 3

know v as a function of s. This may be obtained from the equation of motion. Using Newton's laws of motion, resolving along the tangent to the curve at P, and noting that the reaction has no tangential component, we have

$$m \frac{\mathrm{d}v}{\mathrm{d}t} = mg \sin \psi.$$

This may be written

$$\frac{v \, \mathrm{d}v}{\mathrm{d}s} = g \sin \psi$$

or

$$v \, \mathrm{d}v = g \sin \psi \, \mathrm{d}s.$$

Integrating

$$[\tfrac{1}{2}v^2]_A^P = \int_A^P g \sin \psi \, \mathrm{d}s.$$

The right-hand side is a line integral, the path of integration being the portion AP of the cycloid. Substituting

$$\sin \psi = s/4a, \quad \text{we obtain}$$

$$\tfrac{1}{2}v^2 = \int_{2a}^s \frac{g}{4a} s \, \mathrm{d}s = \frac{g}{8a}(s^2 - 4a^2).$$

Thus
$$v = \sqrt{\left[\frac{g}{4a}(s^2 - 4a^2)\right]}.$$

Hence
$$T = \sqrt{\left(\frac{4a}{g}\right)} \int_{2a}^{4a} \frac{ds}{\sqrt{(s^2 - 4a^2)}}$$
$$= \sqrt{\left(\frac{4a}{g}\right)} \left[\cosh^{-1}\frac{s}{2a}\right]_{2a}^{4a} = \sqrt{\left(\frac{4a}{g}\right)}(\cosh^{-1} 2 - \cosh^{-1} 1)$$
$$= \sqrt{\left(\frac{4a}{g}\right)} \log(2 + \sqrt{3}).$$

Example 3.

If
$$\mathbf{A} = y\mathbf{i} - z\mathbf{j} + x\mathbf{k}$$
find

(a) $\int_C \mathbf{A}\, ds$ (b) $\int_C \mathbf{A}.d\mathbf{s}$ (c) $\int_C \mathbf{A} \times d\mathbf{s}$

where C is the curve whose equation is given in parametric form by the relation
$$\mathbf{r} = a \cos u\, \mathbf{i} + a \sin u\, \mathbf{j} + au \cot \alpha\, \mathbf{k}$$
where a, α are constants, and the parameter u takes values from 0 to 2π.

$d\mathbf{s} = d\mathbf{r} = (-a \sin u\, \mathbf{i} + a \cos u\, \mathbf{j} + a \cot \alpha\, \mathbf{k})\, du$

$d\mathbf{s}^2 = (a^2 \sin^2 u + a^2 \cos^2 u + a^2 \cot^2 \alpha)\, du^2 = a\, \text{cosec}^2\alpha\, du^2$

giving

$ds = a\, \text{cosec}\, \alpha\, du.$

(a) $\displaystyle\int_C \mathbf{A}\,ds = \int_0^{2\pi}(a\sin u\mathbf{i} - au\cot\alpha\mathbf{j} + a\cos u\mathbf{k})a\,\text{cosec}\,\alpha\,du$

$\qquad = a^2\,\text{cosec}\,\alpha[-\cos u\mathbf{i} - \tfrac{1}{2}u^2\cot\alpha\mathbf{j} + \sin u\mathbf{k}]_0^{2\pi}$

$\qquad = -2\pi^2 a^2\,\text{cosec}\,\alpha\cot\alpha\mathbf{j}.$

(b) $\displaystyle\int_C \mathbf{A}.d\mathbf{s} = \int_0^{2\pi}[-a^2\sin^2 u - a^2 u\cot\alpha\cos u$

$\qquad\qquad\qquad + a^2\cos u\cot\alpha]\,du$

$\qquad = a^2[-\tfrac{1}{2}(u - \tfrac{1}{2}\sin 2u)$

$\qquad\qquad - \cot\alpha(u\sin u + \cos u) + \cot\alpha\sin u]_0^{2\pi}$

$\qquad = -\pi a^2.$

(c) $\displaystyle\int_C \mathbf{A}\times d\mathbf{s} = a^2\int_0^{2\pi}(\sin u\mathbf{i} - u\cot\alpha\mathbf{j} + \cos u\mathbf{k})$

$\qquad\qquad\qquad \times(-\sin u\mathbf{i} + \cos u\mathbf{j} + \cot\alpha\mathbf{k})\,du$

$\qquad = a^2\int_0^{2\pi}[(-u\cot^2\alpha - \cos^2 u)\mathbf{i}$

$\qquad\qquad + (-\sin u\cos u - \sin u\cot\alpha)\mathbf{j}$

$\qquad\qquad + (\sin u\cos u - u\sin u\cot\alpha)\mathbf{k}]\,du$

$\qquad = a^2[-\pi(1 + 2\pi\cot^2\alpha)\mathbf{i} + 2\pi\cot\alpha\,\mathbf{k}].$

4.2. Line integral of grad ϕ

The line integral of a vector field $\mathbf{A}$ over a path C, namely $\displaystyle\int \mathbf{A}.d\mathbf{s}$, generally depends on the two end points of C as well as on the particular path joining the two end points. If however $\mathbf{A}$ is expressible as the gradient of a scalar function ϕ, that is, when

$$\mathbf{A} = \mathbf{G} = \text{grad}\,\phi,$$

$\int \mathbf{G} \cdot \mathrm{d}\mathbf{s}$ depends only on the two end points and is independent of the path joining the points.

This may be seen by noting that

$$\mathbf{G} \cdot \mathrm{d}\mathbf{s} = G_s \, \mathrm{d}s = \frac{\mathrm{d}\phi}{\mathrm{d}s} \, \mathrm{d}s = \mathrm{d}\phi \tag{14}$$

showing that $\mathbf{G} \cdot \mathrm{d}\mathbf{s} = G_x \, \mathrm{d}x + G_y \, \mathrm{d}y + G_z \, \mathrm{d}z$ is a perfect differential. Hence

$$\int_C \mathbf{G} \cdot \mathrm{d}\mathbf{s} = \int_C \mathrm{d}\phi = [\phi]_A^B = \phi(B) - \phi(A). \tag{15}$$

Thus we see that the integral depends only on the values of ϕ at the two end points A and B and not on the particular path.

This result may also be expressed in the following form:

$$\int_C \mathbf{G} \cdot \mathrm{d}\mathbf{s} = \int_{C'} \mathbf{G} \cdot \mathrm{d}\mathbf{s} \tag{16}$$

where C and C' are both paths starting at A and ending at B (Fig. 4). Hence

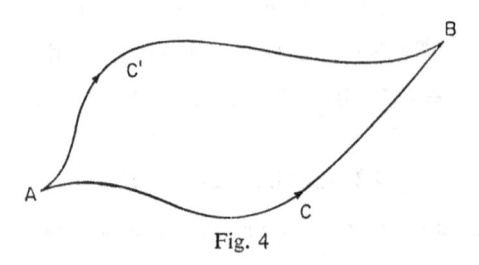

Fig. 4

$$\int_C \mathbf{G} \cdot \mathrm{d}\mathbf{s} + \int_{-C'} \mathbf{G} \cdot \mathrm{d}\mathbf{s} = 0,$$

where $-C'$ denotes that the path of integration is C' taken in the sense B to A.

VECTOR CALCULUS 67

Therefore

$$\oint_{C-C'} \mathbf{G}.d\mathbf{s} = 0$$

where $\oint$ indicates that integration is over the closed curve C–C'. This result being true for all curves C and C', it may be deduced that

$$\oint \mathbf{G}.d\mathbf{s} = 0 \qquad (17)$$

for every closed curve in the region of space under consideration.

The converse theorem also is true, namely that if a vector $\mathbf{A}$ is such that its integral round every closed curve in the region under consideration vanishes, then there exists a point function ϕ such that $\mathbf{A} = \operatorname{grad} \phi$.

Let P_0 be a fixed point in the region and P a variable point. Consider

$$\phi(P) = \int_{P_0}^{P} \mathbf{A}.d\mathbf{s}. \qquad (18)$$

This does not depend on the path joining P_0 and P. Therefore if (x, y, z) are the coordinates of P,

$$\phi = \phi(x, y, z)$$

and is a single valued function of position in the region. Let P' be a point in the neighbourhood of P and let $PP' = s$. Using (18) for the path P_0P'

$$\phi(P') = \int_{P_0}^{P'} \mathbf{A}.d\mathbf{s}. \qquad (19)$$

If the integral (19) is considered with the path P_0P' passing through P, it follows that

$$\phi(P') - \phi(P) = \int_{P}^{P'} \mathbf{A}.d\mathbf{s} \qquad (20)$$

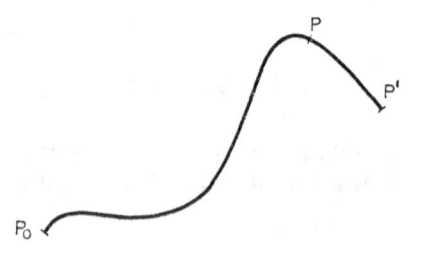

Fig. 5

P and P' being neighbouring points, the right-hand side is approximately $A_s \delta s$ where A_s is the component of **A** in the direction PP'. Hence

$$\phi(P') - \phi(P) = A_s \delta s$$

and we deduce that

$$A_s = \frac{\partial \phi}{\partial s} \qquad (21)$$

This is true for all small displacements from P, and hence **A** = grad ϕ.

A vector **A** with the properties described above is called a *potential vector*. In terms of the Cartesian components A_x, A_y, A_z the condition that $A_x \, dx + A_y \, dy + A_z \, dz$ is a perfect differential may be shown to be

$$\frac{\partial A_y}{\partial z} - \frac{\partial A_z}{\partial y} = 0, \qquad \frac{\partial A_z}{\partial x} - \frac{\partial A_x}{\partial z} = 0, \qquad \frac{\partial A_x}{\partial y} - \frac{\partial A_y}{\partial x} = 0. \qquad (22)$$

As an example of a vector field which satisfies the requirements of a potential vector, consider the work done by a conservative force. If X, Y, Z are the components of the force **F** acting on a particle, the work done by the force as the particle undergoes a displacement from A to B is

$$W = \int_C (X \, dx + Y \, dy + Z \, dz) \qquad (23)$$

where C is the path of the particle from A to B.

If the system is conservative, the work done should not be different for different paths, for if these were different, it would be possible to gain energy by proceeding from A to B by one path and returning from B to A by another path. The conservative nature of the forces requires that the work done in a displacement from A to B, as given by (23), does not depend on the path but only on the end points A and B.

It follows that X, Y, Z should be such that $X\,dx + Y\,dy + Z\,dz$ is a perfect differential, which is written $-dV$ where V is defined as the potential energy of the particle. Thus

$$W = \int_A^B -dV = V(A) - V(B) \tag{24}$$

that is, the work done is equal to the loss of potential energy.

If P_0 is a fixed point taken as the standard configuration corresponding to zero potential energy, and P a variable point whose coordinates are (x, y, z), then

$$V(x, y, z) = -\int_{P_0}^{P} (X\,dx + Y\,dy + Z\,dz) \tag{25}$$

and

$$X = -\frac{\partial V}{\partial x}, \quad Y = -\frac{\partial V}{\partial y}, \quad Z = -\frac{\partial V}{\partial z}. \tag{26}$$

i.e. $\mathbf{F} = -\operatorname{grad} V$

Frequently, P_0 is taken to be at infinity in which case V vanishes at infinity.

Example 1. The gravitational force on a particle of mass m near the earth's surface is represented approximately by the vector

$$-mg\mathbf{k}$$

where the z-axis is vertically upwards and g is a constant. Show that the work done by the force as the particle is displaced from height h_1 to height h_2 along any path is

$$mg(h_1 - h_2).$$

Equation (23) gives

$$W = \int (X\,dx + Y\,dy + Z\,dz) = \int_{h_1}^{h_2} -mg\,dz = -mg[z]_{h_1}^{h_2}$$
$$= mg(h_1 - h_2).$$

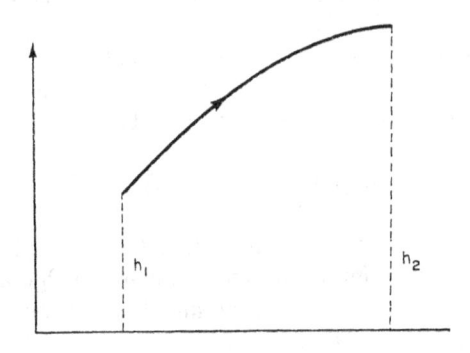

Fig. 6

Example 2. Examine if **A** is a potential vector in the following cases:

(a) $\mathbf{A} = \mathbf{r}f(r)$ (b) $\mathbf{A} = \mathbf{c}f(r)$

where **r** is the position vector, $r = |\mathbf{r}|$ and **c** is a constant vector. Consider $A_x\,dx + A_y\,dy + A_z\,dz$. In case (a),

$$A_x\,dx + A_y\,dy + A_z\,dz = f(r)(x\,dx + y\,dy + z\,dz)$$
$$= f(r)r\,dr = d\phi$$

where $\phi = \int rf(r)\,dr$. Hence **A** is a potential vector. In case (b),

$$A_x\,dx + A_y\,dy + A_z\,dz = f(r)(c_x\,dx + c_y\,dy + c_z\,dz)$$

which is not a perfect differential. It may be verified that conditions (22) are not satisfied. **A** is not a potential vector in this case.

The integral of a vector **A** along a closed curve C, that is, $\int_C \mathbf{A}\cdot d\mathbf{s}$ is called the *circulation* of **A** round the closed curve C.

VECTOR CALCULUS 71

4.3. Surface integrals

The surface integral of a function ϕ over a surface S (which need not be a closed surface or a plane surface) may be defined in a way analogous to the line integral.

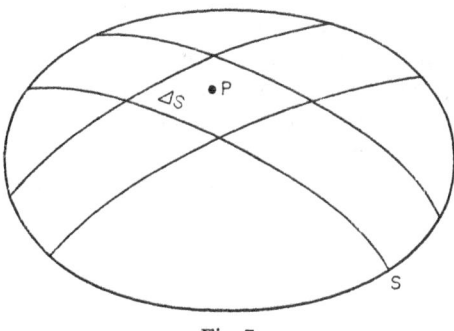

Fig. 7

The surface is divided into m elements ΔS_1, ΔS_2, ... ΔS_m and the expression

$$\phi_1 \Delta S_1 + \phi_2 \Delta S_2 + \ldots + \phi_m \Delta S_m \qquad (27)$$

is formed where $\phi_1, \phi_2, \ldots \phi_m$ are values of ϕ at points $P_1, P_2, \ldots P_m$ which are respectively within each element of surface $\Delta S_1, \Delta S_2, \ldots, \Delta S_m$. If as $m \to \infty$ and each $\Delta S_q \to 0$ for each $q = 1, 2, \ldots, m$, the expression (27) tends to a finite limit, independent of the mode of division into elements, this limit is called the surface integral of ϕ over S and is written

$$\int \phi \, dS. \qquad (28)$$

This definition applies to both vector and scalar integrands.

The coordinates of a point on a surface may be expressed in terms of two parameters which may be written as u and v. For example, a point P on a sphere of radius a may be labelled by the parameters θ and ϕ where θ, ϕ are the angles shown in Fig. 8.

θ is the angle between the axis Oz and the radius vector OP. ϕ is the angle between the planes zOx and zOP. The coordinates of P may be expressed in terms of θ and ϕ by the equations

$$x = a \sin \theta \cos \phi, \qquad y = a \sin \theta \sin \phi, \qquad z = a \cos \theta. \qquad (29)$$

The points on the surface for which $\theta = $ constant will lie on a small circle on the sphere, and points for which $\phi = $ constant will lie on a great circle of which Oz is a diameter. Thus $\theta = $ constant will define a family of curves on the surface, and $\phi = $ constant another family of curves on the surface. Through any point on the surface there passes one curve of the family $\theta = $ constant, and one of the family $\phi = $ constant.

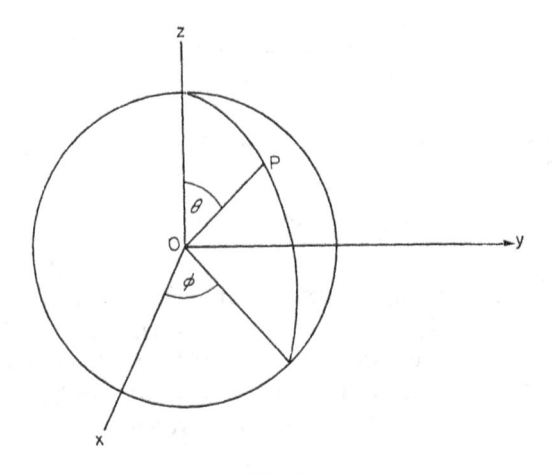

Fig. 8

It may be noted that a surface integral may be expressed as a double integral with respect to the two parameters. Considering again the example of the sphere, suppose we draw the curves $\theta = $ constant, $\phi = $ constant through the point P whose parameters are θ, ϕ, and through a neighbouring point Q whose

VECTOR CALCULUS

parameters are $\theta + \delta\theta$, $\phi + \delta\phi$. Let δS be the area of the element $PRQS$ thus formed (Fig. 9).

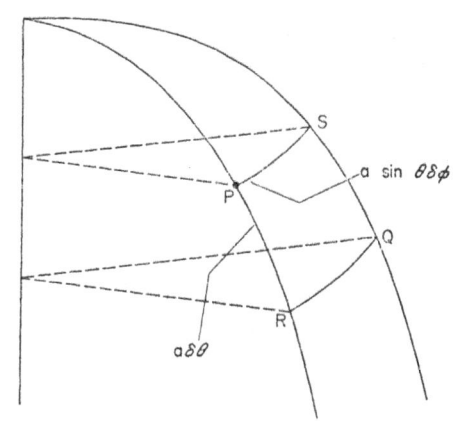

Fig. 9

Approximately $PR = a\delta\theta$, $PS = a\sin\theta\delta\phi$ and $\delta S = (a\delta\theta)(a\sin\theta\delta\phi)$. Hence we take

$$dS = a^2 \sin\theta \, d\theta \, d\phi. \tag{30}$$

Integration over the surface S gives rise to a double integral with respect to θ and ϕ.

Example. Evaluate

$$\text{(i)} \int dS \qquad \text{(ii)} \int z \, dS$$

over the surface of the hemisphere

$$x^2 + y^2 + z^2 = a^2, \qquad z \geqslant 0.$$

Using the expression (30) for dS,

$$\int dS = \int_{\theta=0}^{\pi/2} \int_{\phi=0}^{2\pi} a^2 \sin\theta \, d\theta \, d\phi = 2\pi a^2.$$

The limits of integration have been taken such that as θ, ϕ go through their respective ranges the point P will cover the entire surface. When $\phi =$ constant and θ varies from 0 to $\pi/2$ the point P describes a quadrant. As ϕ varies from 0 to 2π the quadrant generates the hemispherical surface.

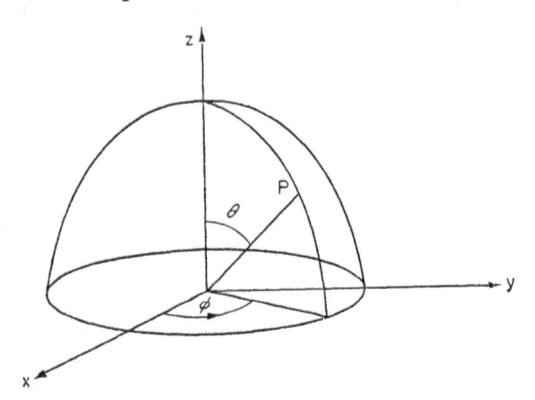

Fig. 10

$$\int z \, dS = \int_0^{\pi/2} \int_0^{2\pi} a \cos\theta \cdot a^2 \sin\theta \, d\theta \, d\phi$$
$$= 2\pi a^3 \int_0^{\pi/2} \sin\theta \cos\theta \, d\theta = \pi a^3.$$

It is useful to define a vector element of area which is written d**S**. Suppose at a point P on the surface the unit vector along the outward normal to the surface is **n**. If dS is the element of area at P, the vector element of area is defined as

$$d\mathbf{S} = \mathbf{n} \, dS, \tag{31}$$

following on the earlier representation of a plane area by a vector normal to the surface.

When the equation of the surface is given in the form

$$f(x, y, z) = \text{constant}, \tag{32}$$

VECTOR CALCULUS 75

the direction cosines of the normal at $P(x, y, z)$ are proportional to $\partial f/\partial x$, $\partial f/\partial y$, $\partial f/\partial z$. Hence

$$\mathbf{n} = \left[\mathbf{i}\frac{\partial f}{\partial x} + \mathbf{j}\frac{\partial f}{\partial y} + \mathbf{k}\frac{\partial f}{\partial z}\right] \bigg/ \sqrt{\left\{\left(\frac{\partial f}{\partial x}\right)^2 + \left(\frac{\partial f}{\partial y}\right)^2 + \left(\frac{\partial f}{\partial z}\right)^2\right\}} \quad (33)$$

$$= \operatorname{grad} f / |\operatorname{grad} f|. \quad (34)$$

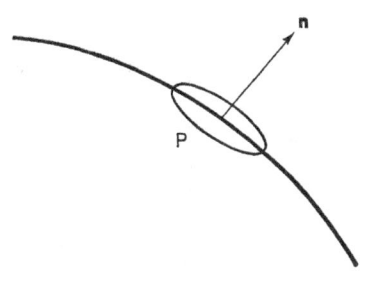

Fig. 11

If the equation of the surface is given in terms of two parameters u and v by relations of the form

$$x = x(u, v), \qquad y = y(u, v), \qquad z = z(u, v), \quad (35)$$

$u = $ const. is the equation of a family of curves on the surface and $v = $ const. is that of another family of curves on the surface (Fig. 12). $(\partial x/\partial u, \partial y/\partial u, \partial z/\partial u)$ are proportional to the direction cosines of the tangent to the curve $v = $ const. in the direction of u increasing, and define a vector. We write this vector as $\mathbf{l}$ where

$$\mathbf{l} = \left(\frac{\partial x}{\partial u}, \frac{\partial y}{\partial u}, \frac{\partial z}{\partial u}\right). \quad (36)$$

Similarly a vector

$$\mathbf{m} = \left(\frac{\partial x}{\partial v}, \frac{\partial y}{\partial v}, \frac{\partial z}{\partial v}\right)$$

has its direction along the tangent to the curve $u = $ const. The unit vector normal to the surface is

$$\mathbf{n} = \frac{\mathbf{l} \times \mathbf{m}}{|\mathbf{l} \times \mathbf{m}|}. \qquad (37)$$

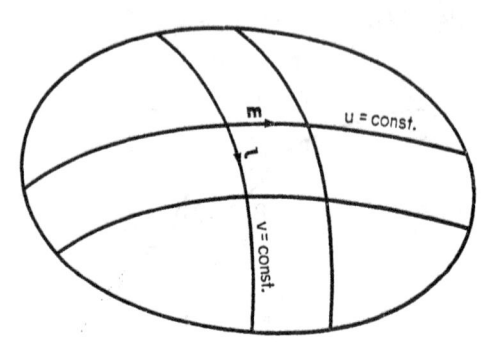

Fig. 12

We may also note by considering the element of area that

$$d\mathbf{S} = \mathbf{l} \times \mathbf{m} \, du \, dv. \qquad (38)$$

In the case of the sphere, taking u as θ and v as ϕ, eqn. (29) gives

$$\mathbf{l} = a \cos \theta \cos \phi \mathbf{i} + a \cos \theta \sin \phi \mathbf{j} - a \sin \theta \mathbf{k}$$

$$\mathbf{m} = -a \sin \theta \sin \phi \mathbf{i} + a \sin \theta \cos \phi \mathbf{j}$$

$$\mathbf{l} \times \mathbf{m} = a^2 \sin^2 \theta \cos \phi \mathbf{i} + a^2 \sin^2 \theta \sin \phi \mathbf{i} + a \sin \theta \cos \theta \mathbf{k}$$

$$= a^2 \sin \theta [\sin \theta \cos \phi \mathbf{i} + \sin \theta \sin \phi \mathbf{j} + \cos \theta \mathbf{k}]$$

$$= a^2 \sin \theta \, \mathbf{n}.$$

We verify that (38) gives

$$d\mathbf{S} = a^2 \sin \theta \, d\theta \, d\phi \, \mathbf{n}$$

which agrees with the earlier result (30).

VECTOR CALCULUS 77

By the surface integral of a vector **A** over a surface S is meant

$$\int_S \mathbf{A} \cdot d\mathbf{S} \qquad (39)$$

which may also be written as

$$\int_S \mathbf{A} \cdot \mathbf{n}\, dS \quad \text{or} \quad \int_S A_n\, dS \qquad (40)$$

where A_n denotes the component of **A** in the direction of the outward normal at the point concerned. This integral is called the flux of **A** out of the surface **S**.

Example. Evaluate $\int_S \mathbf{A} \cdot d\mathbf{S}$ where

$$\mathbf{A} = \frac{e}{r^3} \mathbf{r}$$

and S is the curved surface of the cylinder $x^2 + y^2 = a^2$ from $z = 0$ to $z = h$.

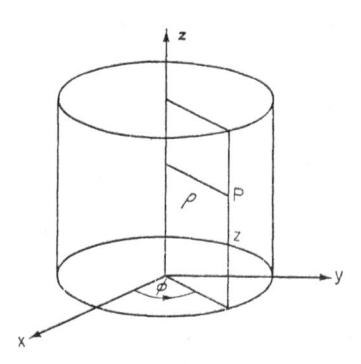

Fig. 13

Here it is convenient to take cylindrical polar coordinates (ρ, ϕ, z), as in Fig. 13, such that

$$x = \rho \cos \phi, \qquad y = \rho \sin \phi.$$

On the surface,

$$\rho = a, \quad 0 \leqslant \phi \leqslant 2\pi, \quad 0 \leqslant z \leqslant h.$$

$$\mathbf{A} = \frac{e}{r^3}\mathbf{r} = \frac{e}{(a^2+z^2)^{3/2}}[a\cos\phi\mathbf{i} + a\sin\phi\mathbf{j} + z\mathbf{k}]$$

$$\mathbf{n} = \cos\phi\mathbf{i} + \sin\phi\mathbf{j}$$

$$A_n = \frac{e}{(a^2+z^2)^{3/2}}(a\cos^2\phi + a\sin^2\phi) = \frac{ea}{(a^2+z^2)^{3/2}}.$$

$$dS = a\,d\phi\,dz.$$

$$\int_S \mathbf{A}\cdot d\mathbf{S} = \int_{z=0}^h \int_{\phi=0}^{2\pi} \frac{ea^2}{(a^2+z^2)^{3/2}}\,dz\,d\phi$$

$$= 2\pi ea^2 \int_0^h \frac{dz}{(a^2+z^2)^{3/2}} = \frac{2\pi eh}{\sqrt{(a^2+h^2)}}.$$

4.4. Volume integrals

A volume integral may be defined in the same way as line and surface integrals, that is, a given volume Ω is divided into elements, of volume $\Delta\Omega_1, \Delta\Omega_2, \ldots, \Delta\Omega_m$ respectively and the expression

$$\phi_1\Delta\Omega_1 + \phi_2\Delta\Omega_2 + \ldots + \phi_m\Delta\Omega_m \tag{41}$$

is formed, where ϕ_q denotes the value of ϕ at a point within the volume $\Delta\Omega_q$, $q = 1, 2, \ldots, m$. If as $m \to \infty$ and $\Delta\Omega_q \to 0$ for each q, the expression (41) tends to a finite limit, independent of the mode of division into elements, that limit is called the volume integral of ϕ over the volume Ω and is written

$$\int_\Omega \phi\,d\Omega. \tag{42}$$

This definition applies to both vector and scalar integrands.

A volume integral is equivalent to a triple integral. In Cartesian

VECTOR CALCULUS 79

coordinates if we take as infinitesimal element of volume a rectangular solid with sides dx, dy, dz parallel to the axes respectively

$$d\Omega = dx\, dy\, dz \tag{43}$$

In general, we may use three parameters u, v, w say, to express the coordinates of a point in the form

$$x = x(u, v, w), \qquad y = y(u, v, w), \qquad z = z(u, v, w). \tag{44}$$

The volume integral will then be a triple integral with respect to the three parameters u, v, w.

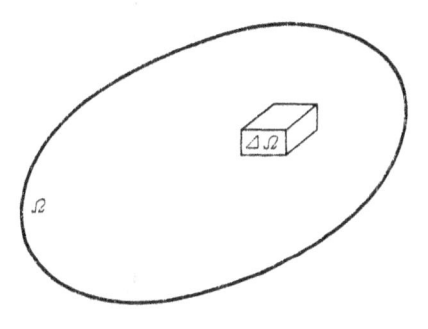

Fig. 14

In spherical polar coordinates, r, θ, ϕ, where r is the radial distance and θ, ϕ are the angles shown earlier in Fig. 8, may be taken as parameters. The equations connecting x, y, z, and r, θ, ϕ are

$$x = r \sin \theta \cos \phi, \qquad y = r \sin \theta \sin \phi, \qquad z = r \cos \theta. \tag{45}$$

If we draw the surfaces $r =$ const., $\theta =$ const., $\phi =$ const. through a point whose parameters are r, θ, ϕ and through a neighbouring point whose parameters are $r + \delta r$, $\theta + \delta\theta$, $\phi + \delta\phi$, we obtain a small box whose sides are approximately δr, $r\delta\theta$, $r \sin \theta \delta\phi$ (Fig. 15). Hence

$$d\Omega = r^2 \sin \theta\, dr\, d\theta\, d\phi. \tag{46}$$

If we use cylindrical polar coordinates ρ, ϕ, z where

$$x = \rho \cos \phi, \qquad y = \rho \sin \phi,$$

then

$$d\Omega = \rho \, d\rho \, d\phi \, dz. \tag{47}$$

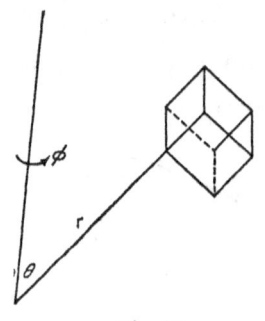

Fig. 15

In general when the coordinates are given in parametric form by equations such as (44), we state (without giving the proof) that

$$d\Omega = J \, du \, dv \, dw \tag{48}$$

where

$$J = \begin{vmatrix} \dfrac{\partial x}{\partial u} & \dfrac{\partial y}{\partial u} & \dfrac{\partial z}{\partial u} \\ \dfrac{\partial x}{\partial v} & \dfrac{\partial y}{\partial v} & \dfrac{\partial z}{\partial v} \\ \dfrac{\partial x}{\partial w} & \dfrac{\partial y}{\partial w} & \dfrac{\partial z}{\partial w} \end{vmatrix} \tag{49}$$

is called the *Jacobian* of the transformation from x, y, z to u, v, w. We may verify that (48) agrees with (46) and (47) in these particular cases.

Example. Find the volume of the ellipsoid

$$\frac{x^2}{a^2} + \frac{y^2}{b^2} + \frac{z^2}{c^2} = 1.$$

Introduce a transformation of coordinates from x, y, z to X, Y, Z where

$$x = aX, \qquad y = bY, \qquad z = cZ.$$

Then $\mathrm{d}x\,\mathrm{d}y\,\mathrm{d}z = abc\,\mathrm{d}X\,\mathrm{d}Y\,\mathrm{d}Z$.

The surface of the ellipsoid will transform to the sphere $X^2 + Y^2 + Z^2 = 1$ in the new coordinate system. Hence the volume of the ellipsoid is

$$\iiint\limits_{ellipsoid} \mathrm{d}x\,\mathrm{d}y\,\mathrm{d}z = abc \iiint\limits_{unit\ sphere} \mathrm{d}X\,\mathrm{d}Y\,\mathrm{d}Z.$$

The integral on the right-hand side is just the volume of the unit sphere and is therefore $(4/3)\pi$. Hence the required volume is $(4/3)\pi abc$.

As an application in the use of spherical polar coordinates it would be of interest to derive the expression $(4/3)\pi$ for the volume of a sphere of unit radius.

Using (46), this volume is

$$\int_{r=0}^{1} \int_{\theta=0}^{\pi} \int_{\phi=0}^{2\pi} r^2 \sin\theta\,\mathrm{d}r\,\mathrm{d}\theta\,\mathrm{d}\phi$$
$$= 2\pi \int_0^1 \int_0^\pi r^2 \sin\theta\,\mathrm{d}r\,\mathrm{d}\theta = 2\pi \int_0^1 r^2\,\mathrm{d}r[-\cos\theta]_0^\pi$$
$$= 2\pi.2. \int_0^1 r^2\,\mathrm{d}r = \tfrac{4}{3}\pi.$$

4.5. Divergence

Given a vector field **A**, its divergence at any point P is defined as follows.

Let $\Delta\Omega$ be a small volume surrounding the point P, and whose bounding surface is ΔS.

$\int_{\Delta S} A_n \, dS$ is the flux of **A** out of the surface ΔS. Consider the ratio of this flux to the volume enclosed by the surface, that is, consider the expression

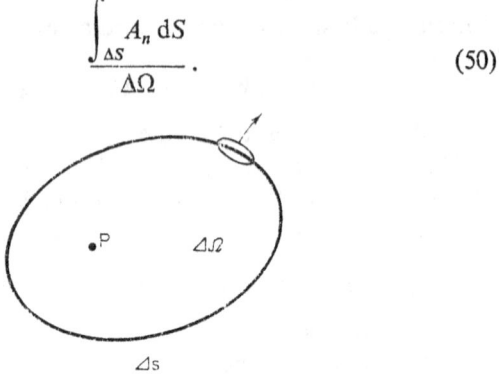

$$\frac{\int_{\Delta S} A_n \, dS}{\Delta \Omega}. \tag{50}$$

Fig. 16

If **A** is such that, as the volume $\Delta \Omega \to 0$ still enclosing the point P, the expression (50) tends to a finite limit, independent of the shape of the surface, this limit is called the *divergence of* **A** *at the point P* and is written div **A**. It may be noted that div **A** is a scalar quantity and represents the flux of **A** per unit volume at P; div **A** thus describes a scalar field.

For example, if the point vector **v** gives the velocity field of a moving liquid and liquid enters the system at a uniform rate from a source situated at a point Q in the liquid, then div **v** at Q is a measure of the strength of the source. At other points where there are no sources or sinks div **v** $= 0$.

div **A** *in Cartesian Coordinates*

A rectangular box of sides δx, δy, δz parallel to the axes is constructed with the point $P(x, y, z)$ as the centre, and the definition given above is applied with this box to determine div **A** at P. In evaluating the numerator of the expression (50) it may be noted that the solid has six faces. It is convenient to group them in pairs of parallel faces. Considering the contribution from the face *BCGF*, noting that the area of the face is $\delta y \delta z$, we obtain the flux as approximately

$$A_x(x + \tfrac{1}{2}\delta x, y, z)\delta y \delta z.$$

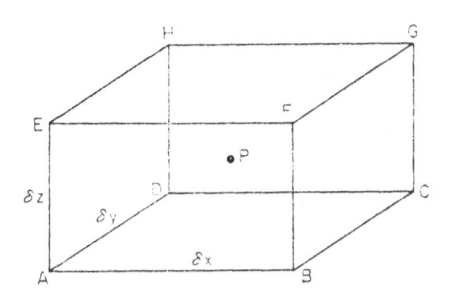

Fig. 17

The contribution from the face *ADHE* is

$$-A_x(x - \tfrac{1}{2}\delta x, y, z)\delta y \delta z,$$

the minus sign arising because the outward normal to this face is in the negative direction of the x-axis. Taking the contributions from these two parallel faces together, we obtain

$$[A_x(x + \tfrac{1}{2}\delta x, y, z) - A_x(x - \tfrac{1}{2}\delta x, y, z)]\delta y \delta z$$

which is approximately

$$\frac{\partial A_x}{\partial x}\delta x \delta y \delta z.$$

In the same way the contributions to the flux from the other two pairs of faces are

$$\frac{\partial A_y}{\partial y}\delta x \delta y \delta z \quad \text{and} \quad \frac{\partial A_z}{\partial z}\delta x \delta y \delta z,$$

thus making the total flux as approximately

$$\left(\frac{\partial A_x}{\partial x} + \frac{\partial A_y}{\partial y} + \frac{\partial A_z}{\partial z}\right)\delta x \delta y \delta z.$$

Also $\Delta\Omega = \delta x \delta y \delta z$. Hence, forming the expression (50) and proceeding to the limit we have

$$\text{div } \mathbf{A} = \frac{\partial A_x}{\partial x} + \frac{\partial A_y}{\partial y} + \frac{\partial A_z}{\partial z}. \tag{51}$$

For example, if $\mathbf{A} = \mathbf{r} = x\mathbf{i} + y\mathbf{j} + z\mathbf{k}$,

$$\text{div } \mathbf{A} = \frac{\partial}{\partial x}(x) + \frac{\partial}{\partial y}(y) + \frac{\partial}{\partial z}(z) = 3.$$

If $\mathbf{A} = \text{grad } \phi$,

$$\text{div (grad } \phi) = \frac{\partial}{\partial x}\left(\frac{\partial \phi}{\partial x}\right) + \frac{\partial}{\partial y}\left(\frac{\partial \phi}{\partial y}\right) + \frac{\partial}{\partial z}\left(\frac{\partial \phi}{\partial z}\right)$$

$$= \frac{\partial^2 \phi}{\partial x^2} + \frac{\partial^2 \phi}{\partial y^2} + \frac{\partial^2 \phi}{\partial z^2}.$$

If

$$\mathbf{A} = -\frac{y}{x^2 + y^2}\mathbf{i} + \frac{x}{x^2 + y^2}\mathbf{j},$$

$$\text{div } \mathbf{A} = \frac{\partial}{\partial x}\left(-\frac{y}{x^2 + y^2}\right) + \frac{\partial}{\partial y}\left(\frac{x}{x^2 + y^2}\right)$$

$$= (-y)\frac{(-1)(2x)}{(x^2 + y^2)^2} + (x)\frac{(-1)(2y)}{(x^2 + y^2)^2} = 0.$$

A vector whose divergence vanishes everywhere is said to be *solenoidal*. These vectors are useful in various branches of physics.

The equations

$$\text{div}(\mathbf{A} + \mathbf{B}) = \text{div } \mathbf{A} + \text{div } \mathbf{B}$$

$$\text{div}(\lambda \mathbf{A}) = \lambda \text{ div } \mathbf{A}$$

VECTOR CALCULUS 85

where λ is a constant scalar, may be easily obtained from the definition.

If ϕ is a scalar point function and $\mathbf{A}$ a vector point function, it may be shown that

$$\operatorname{div}(\phi \mathbf{A}) = \phi \operatorname{div} \mathbf{A} + (\operatorname{grad} \phi) . \mathbf{A} \tag{52}$$

If we use Cartesian coordinates,

$$\begin{aligned}
\text{L.H.S.} &= \frac{\partial}{\partial x}(\phi A_x) + \frac{\partial}{\partial y}(\phi A_y) + \frac{\partial}{\partial z}(\phi A_z) \\
&= \phi \frac{\partial A_x}{\partial x} + \frac{\partial \phi}{\partial x} A_x + \phi \frac{\partial A_y}{\partial y} + \frac{\partial \phi}{\partial y} A_y + \phi \frac{\partial A_z}{\partial z} + \frac{\partial \phi}{\partial z} A_z \\
&= \phi \left(\frac{\partial A_x}{\partial x} + \frac{\partial A_y}{\partial y} + \frac{\partial A_z}{\partial z} \right) + \left(\frac{\partial \phi}{\partial x} A_x + \frac{\partial \phi}{\partial y} A_y + \frac{\partial \phi}{\partial z} A_z \right) \\
&= \text{R.H.S.}
\end{aligned}$$

We have shown the result for Cartesian coordinates, but it is true in any coordinate system.

4.6. Gauss's transformation

If S is a closed surface enclosing a volume Ω of space and $\mathbf{A}$ is a vector field,

$$\int_S \mathbf{A} . d\mathbf{S} = \int_\Omega \operatorname{div} \mathbf{A} \, d\Omega. \tag{53}$$

The result may be obtained in the following way. Divide the volume into a number of cells of volume $\Delta\Omega_q$ and surface area ΔS_q, $q = 1, 2, \ldots, m$. From the definition of divergence as the limit of the expression (50), it may be written for each q

$$\int_{\Delta S_q} A_n \, dS = \operatorname{div} A \, d\Omega_q + \varepsilon_q \Delta\Omega_q \tag{54}$$

where $\varepsilon_q \to 0$ as $\Delta\Omega_q \to 0$. Adding all these equations for $q = 1, 2, \ldots, m$, noting that the contributions to the flux from adjacent faces of neighbouring cells cancel out in the summation, and proceeding to the limit $\Delta\Omega_q \to 0$ for each q and $m \to \infty$, the result (53) is obtained.

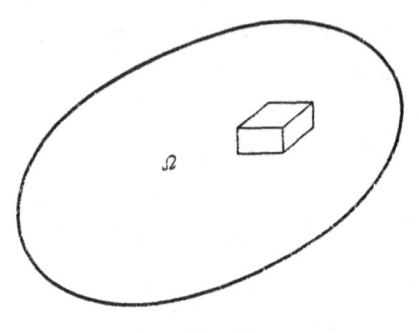

Fig. 18

Example 1. Verify the equality of the surface and volume integrals in the eqn. (53) when

$$\mathbf{A} = \mu\mathbf{r}$$

where μ is a constant, $\mathbf{r}$ is the radius vector and the surface of integration is the sphere with centre at the origin and of radius a.

On the spherical surface $|\mathbf{r}| = a$,

$$\mathbf{n} = \mathbf{r}/a$$

giving $A_n = \mu\mathbf{r}\cdot\mathbf{n} = \mu a$.

Hence

$$\int_S A_n \, dS = \mu a \int dS = \mu a \cdot 4\pi a^2 = 4\pi\mu a^3.$$

$$\text{div } \mathbf{A} = 3\mu$$

$$\int_\Omega \text{div } \mathbf{A} \, d\Omega = 3\mu \int_\Omega d\Omega = 3\mu(4/3)\pi a^3 = 4\pi\mu a^3.$$

VECTOR CALCULUS

Example 2. If S *is the surface of the ellipsoid*

$$\frac{x^2}{a^2} + \frac{y^2}{b^2} + \frac{z^2}{c^2} = 1$$

evaluate

$$\int_S \frac{dS}{p}$$

where p is the perpendicular distance from the origin of the tangent plane at the point P(x, y, z) on the ellipsoid.

The perpendicular distance from the origin of the tangent plane is

$$p = 1 \bigg/ \sqrt{\left\{\frac{x^2}{a^4} + \frac{y^2}{b^4} + \frac{z^2}{c^4}\right\}}.$$

The normal at P to the ellipsoid has its direction cosines proportional to

$$\left(\frac{x}{a^2}, \frac{y}{b^2}, \frac{z}{c^2}\right).$$

The unit vector along the normal is thus seen to be

$$\mathbf{n} = \left(\frac{px}{a^2}, \frac{py}{b^2}, \frac{pz}{c^2}\right).$$

Suppose we take

$$\mathbf{A} = \left(\frac{x}{a^2}, \frac{y}{b^2}, \frac{z}{c^2}\right),$$

which we can do because $\mathbf{A}$ so chosen is a vector, namely $\mathbf{A} = (1/p)\mathbf{n}$, and apply the transformation (53).

$$A_n = \frac{px}{a^2} \cdot \frac{x}{a^2} + \frac{py}{b^2} \cdot \frac{y}{b^2} + \frac{pz}{c^2} \cdot \frac{z}{c^2} = \frac{1}{p}$$

Hence the surface integral in (53) is

$$\int \frac{1}{p} \, dS.$$

Now

$$\text{div } \mathbf{A} = \frac{\partial}{\partial x}\left(\frac{x}{a^2}\right) + \frac{\partial}{\partial y}\left(\frac{y}{a^2}\right) + \frac{\partial}{\partial z}\left(\frac{z}{a^2}\right)$$

$$= \frac{1}{a^2} + \frac{1}{b^2} + \frac{1}{c^2}.$$

Hence the volume integral is

$$\left(\frac{1}{a^2} + \frac{1}{b^2} + \frac{1}{c^2}\right)\int d\Omega = \left(\frac{1}{a^2} + \frac{1}{b^2} + \frac{1}{c^2}\right)\tfrac{4}{3}\pi abc.$$

Thus we obtain

$$\int_s \frac{dS}{p} = \tfrac{4}{3}\pi abc\left(\frac{1}{a^2} + \frac{1}{b^2} + \frac{1}{c^2}\right).$$

Here we have determined the surface integral, not by evaluating it directly but by transforming it, by means of Gauss's transformation, into a volume integral which turns out to be simple to evaluate.

4.7. Curl A

Given a vector field $\mathbf{A}$, another vector field curl $\mathbf{A}$ may be derived from it in the following way:

Suppose ΔS is a surface element at a point P, the rim of the element being the closed curve ΔC. Let $\mathbf{n}$ be unit vector along the normal at P to the surface. The line integral of $\mathbf{A}$ over the closed curve ΔC, that is, the circulation of A round ΔC, is

$$\oint_{\Delta C} \mathbf{A} \cdot d\mathbf{s}$$

The expression

$$\frac{\oint_{\Delta C} \mathbf{A}.d\mathbf{s}}{\Delta S} \tag{55}$$

is formed, and its behaviour as $\Delta S \to 0$ is investigated. If the expression (55) tends to a finite limit, independent of the shape of the curve, this limit defines the value of

$$(\operatorname{curl} \mathbf{A})_n \tag{56}$$

which is the component of a certain vector curl $\mathbf{A}$ along the normal $\mathbf{n}$ to the surface. The sense of the normal, and the sense in which the circulation is taken, are related by the right-hand screw rule.

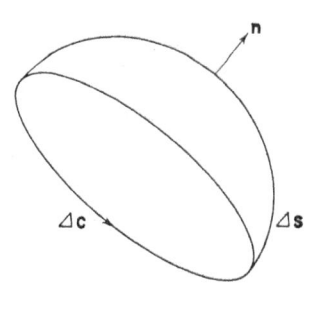

Fig. 19

Curl $\mathbf{A}$ *in Cartesian Components*

To find the component of curl $\mathbf{A}$ in the x-direction, the procedure outlined above for the definitions of $(\operatorname{curl} \mathbf{A})_n$ is applied with the surface S to be a small rectangular area of centre $P(x, y, z)$ and sides parallel to the axes of y and z, and of lengths δy and δz respectively, the x-axis being normal to the plane of the rectangle.

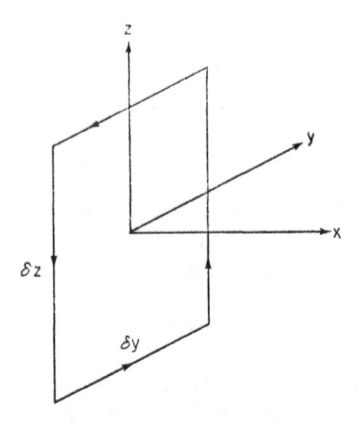

Fig. 20

$\oint \mathbf{A} \cdot d\mathbf{s}$ taken round the perimeter of the rectangle in the sense indicated in Fig. 20 is approximately

$$A_y(x, y, z - \tfrac{1}{2}\delta z)\delta y + A_z(x, y + \tfrac{1}{2}\delta y, z)\delta z$$
$$- A_y(x, y, z + \tfrac{1}{2}\delta z)\delta y - A_z(x, y - \tfrac{1}{2}\delta y, z)\delta z$$
$$= -\frac{\partial A_y}{\partial z}\delta z \delta y + \frac{\partial A_z}{\partial y}\delta y \delta z.$$

$$\Delta S = \delta y \delta z.$$

Hence the limit of the expression (55) gives

$$\left. \begin{aligned} (\operatorname{curl} \mathbf{A})_x &= \frac{\partial A_z}{\partial y} - \frac{\partial A_y}{\partial z}. \\ \text{Similarly} \quad (\operatorname{curl} \mathbf{A})_y &= \frac{\partial A_x}{\partial z} - \frac{\partial A_z}{\partial x} \\ (\operatorname{curl} \mathbf{A})_z &= \frac{\partial A_y}{\partial x} - \frac{\partial A_x}{\partial y} \end{aligned} \right\} \quad (57)$$

Thus

$$\operatorname{curl} \mathbf{A} = \mathbf{i}\left(\frac{\partial A_z}{\partial y} - \frac{\partial A_y}{\partial z}\right) + \mathbf{j}\left(\frac{\partial A_x}{\partial z} - \frac{\partial A_z}{\partial x}\right) + \mathbf{k}\left(\frac{\partial A_y}{\partial x} - \frac{\partial A_x}{\partial y}\right)$$

$$= \begin{vmatrix} \mathbf{i} & \mathbf{j} & \mathbf{k} \\ \dfrac{\partial}{\partial x} & \dfrac{\partial}{\partial y} & \dfrac{\partial}{\partial z} \\ A_x & A_y & A_z \end{vmatrix} \qquad (58)$$

The expression (58) as a determinant is in a form convenient to remember.

Example 1. If $\mathbf{A} = \operatorname{grad} \phi$, find curl $\mathbf{A}$.

$$A_x = \frac{\partial \phi}{\partial x}, \quad A_y = \frac{\partial \phi}{\partial y}, \quad A_z = \frac{\partial \phi}{\partial z}.$$

From (57)

$$(\operatorname{curl} \mathbf{A})_x = \frac{\partial}{\partial y}\left(\frac{\partial \phi}{\partial z}\right) - \frac{\partial}{\partial z}\left(\frac{\partial \phi}{\partial y}\right) = \frac{\partial^2 \phi}{\partial y\, \partial z} - \frac{\partial^2 \phi}{\partial z\, \partial y} = 0.$$

Similarly

$$(\operatorname{curl} \mathbf{A})_y = 0 \quad \text{and} \quad (\operatorname{curl} \mathbf{A})_z = 0.$$

Thus

$$\operatorname{curl}(\operatorname{grad} \phi) \equiv \mathbf{0}.$$

Example 2. If $\mathbf{A} = (-\omega y, \omega x, 0)$ where ω is a constant, find curl $\mathbf{A}$.

$$(\operatorname{curl} \mathbf{A})_x = \frac{\partial}{\partial y}(0) - \frac{\partial}{\partial z}(\omega x) = 0$$

$$(\operatorname{curl} \mathbf{A})_y = \frac{\partial}{\partial z}(-\omega y) - \frac{\partial}{\partial x}(0) = 0$$

$$(\operatorname{curl} \mathbf{A})_z = \frac{\partial}{\partial x}(\omega x) - \frac{\partial}{\partial y}(-\omega y) = 2\omega.$$

Thus curl $\mathbf{A} = (0, 0, 2\omega)$.

If **v** describes the velocity field of a moving liquid, curl **v** is called the *vorticity* of the liquid. If a liquid has no rotatory motion (for example if it is moving uniformly along a canal) then curl **v** = **O**. In the example given above **A** is the velocity of a liquid that is rotating steadily with constant angular velocity about the z-axis. The vorticity vector in this case has no component in the x- or y-direction, but has a component in the z-direction.

4.8. Stokes' theorem

Given a vector field **A** and a surface S whose boundary is a closed curve C, then

$$\oint_C \mathbf{A} \cdot \mathrm{d}\mathbf{s} = \int_S (\operatorname{curl} \mathbf{A})_n \, \mathrm{d}S. \tag{59}$$

Divide S into small areas ΔS_q with perimeters $C_q (q = 1, \ldots, m)$ as in Fig. 21. From the definition of curl A as a limit of the expression (55), we have for each q

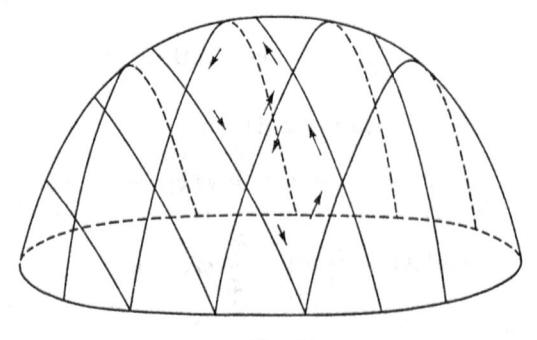

Fig. 21

$$\oint_{C_q} \mathbf{A} \cdot \mathrm{d}\mathbf{s} = (\operatorname{curl} \mathbf{A})_n \Delta S_q + \varepsilon_q \Delta S_q \tag{60}$$

where $\varepsilon_q \to 0$ as $\Delta S_q \to 0$. Adding on all eqn. (60) for $q = 1, 2, \ldots, m$, noting that the contributions to the circulation from adjacent boundaries of neighbouring meshes cancel out, and proceeding to the limits $m \to \infty$, the result (59) is obtained.

Example. *The surface S is the hemisphere*

$$x^2 + y^2 + z^2 = 1, \quad z \geqslant 0.$$

A *is the vector whose components are* (y, z, x). *Verify Stokes' theorem by evaluating each side directly.*

(London S. 1944)

Using spherical polar coordinates

$$\mathbf{n} = \sin\theta\cos\phi\,\mathbf{i} + \sin\theta\sin\phi\,\mathbf{j} + \cos\theta\,\mathbf{k}.$$

Since $\mathbf{A} = (y, z, x)$, and curl $\mathbf{A} = -(\mathbf{i} + \mathbf{j} + \mathbf{k})$

$$(\text{curl }\mathbf{A})_n = -\sin\theta\cos\phi - \sin\theta\sin\phi - \cos\theta.$$

Therefore

$$\int_S (\text{curl }\mathbf{A})_n \, dS = \int_{\theta=0}^{\pi/2} \int_{\phi=0}^{2\pi} [-\sin\theta\cos\phi - \sin\theta\sin\phi - \\ - \cos\theta]\sin\theta \, d\theta \, d\phi$$

$$= -2\pi \int_0^{\pi/2} \sin\theta\cos\theta \, d\theta = -\pi.$$

Again, on the circle $x^2 + y^2 = a^2, z = 0$,

$$d\mathbf{s} = (-a\sin\phi\,\mathbf{i} + a\cos\phi\,\mathbf{j})\,d\phi$$

$$\mathbf{A}.d\mathbf{s} = -\sin^2\phi\,d\phi$$

$$\oint_C \mathbf{A}.d\mathbf{s} = -\int_0^{2\pi} \sin^2\phi\,d\phi = -\pi$$

which verifies the result. Stokes' theorem provides a useful transformation from line integrals to surface integrals, and vice versa, and has many applications.

4.9. The operator ∇

We have seen that the gradient of a scalar point function ϕ may be written in Cartesian coordinates as

$$\operatorname{grad} \phi = \mathbf{i}\frac{\partial \phi}{\partial x} + \mathbf{j}\frac{\partial \phi}{\partial y} + \mathbf{k}\frac{\partial \phi}{\partial z}, \qquad (61)$$

where $\mathbf{i}, \mathbf{j}, \mathbf{k}$ are unit vectors along the three axes respectively. It is convenient to introduce the operator

$$\nabla \equiv \mathbf{i}\frac{\partial}{\partial x} + \mathbf{j}\frac{\partial}{\partial y} + \mathbf{k}\frac{\partial}{\partial z}. \qquad (62)$$

This is called "nabla" or "del" operator.

In terms of this operator ∇,

$$\operatorname{grad} \phi = \nabla \phi. \qquad (63)$$

The introduction of the operator ∇ provides a short hand for writing down certain equations of vector calculus. ∇ is a differential vector operator—differential because it involves the operators of differentiaton, such as $\partial/\partial x$, $\partial/\partial y$, $\partial/\partial z$ and their combinations, and vector because it is linear in $\mathbf{i}, \mathbf{j}, \mathbf{k}$.

In this notation

$$\begin{aligned} \mathrm{d}\phi &= \frac{\partial \phi}{\partial x}\mathrm{d}x + \frac{\partial \phi}{\partial y}\mathrm{d}y + \frac{\partial \phi}{\partial z}\mathrm{d}z \\ &= (\nabla \phi).\mathrm{d}\mathbf{r}. \end{aligned} \qquad (64)$$

Again,

$$\begin{aligned} \nabla . \mathbf{A} &= \left(\mathbf{i}\frac{\partial}{\partial x} + \mathbf{j}\frac{\partial}{\partial y} + \mathbf{k}\frac{\partial}{\partial z}\right).(\mathbf{i}A_x + \mathbf{j}A_y + \mathbf{k}A_z) \\ &= \frac{\partial A_x}{\partial x} + \frac{\partial A_y}{\partial y} + \frac{\partial A_z}{\partial z} = \operatorname{div} \mathbf{A}. \end{aligned}$$

$$\nabla \times \mathbf{A} = \left(\mathbf{i}\frac{\partial}{\partial x} + \mathbf{j}\frac{\partial}{\partial y} + \mathbf{k}\frac{\partial}{\partial z}\right) \times (\mathbf{i}A_x + \mathbf{j}A_y + \mathbf{k}\,A_z)$$

VECTOR CALCULUS 95

$$= \mathbf{i}\left(\frac{\partial A_z}{\partial y} - \frac{\partial A_y}{\partial z}\right) + \mathbf{j}\left(\frac{\partial A_x}{\partial z} - \frac{\partial A_z}{\partial x}\right) + \mathbf{k}\left(\frac{\partial A_y}{\partial x} - \frac{\partial A_x}{\partial y}\right)$$
$$= \text{curl } \mathbf{A}.$$

Hence the three quantities which have been specially studied in vector calculus, namely grad ϕ, div $\mathbf{A}$ and curl $\mathbf{A}$ are expressed simply in terms of the operator ∇ by the relations

$$\left.\begin{aligned}\text{grad } \phi &= \nabla \phi \\ \text{div } \mathbf{A} &= \nabla \cdot \mathbf{A} \\ \text{curl } \mathbf{A} &= \nabla \times \mathbf{A}\end{aligned}\right\} \tag{65}$$

The symbol ∇ may be used in combination with other functions, according to the rules of vector algebra, and yielding new expressions in so far as meanings can be given to these expressions. Thus when combining ∇ and $\mathbf{A}$, $\nabla \times \mathbf{A}$ has a meaning, but $\mathbf{A} \times \nabla$ has no meaning when standing by itself, being an operator which takes on a meaning only when it is combined with an operand. Keeping this point in mind the symbol ∇ may be used with great advantage to simplify complicated expressions of vector calculus.

For example, consider the relation

$$\text{curl}(\text{grad } \phi) \equiv O \tag{66}$$

which was shown in example 1 of §4.7. If we write this symbolically as $\nabla \times (\nabla \phi)$, its vanishing may be inferred from the occurrence of $\nabla \times \nabla$ which vanishes. In the same way, consider div (curl $\mathbf{A}$). This may be written as $\nabla \cdot (\nabla \times \mathbf{A})$ which vanishes since ∇ occurs twice in the triple product. Hence

$$\text{div}(\text{curl } \mathbf{A}) \equiv O. \tag{67}$$

We may prove the following results, using ∇.

$$\text{div}(\phi \mathbf{A}) = \phi \text{ div } \mathbf{A} + (\text{grad } \phi) \cdot \mathbf{A} \tag{68}$$

$$\text{curl } \phi \mathbf{A} = \phi \text{ curl } \mathbf{A} + (\text{grad } \phi) \times \mathbf{A} \tag{69}$$

$$\text{div}(\mathbf{A} \times \mathbf{B}) = \mathbf{B} \cdot \text{curl } \mathbf{A} - \mathbf{A} \cdot \text{curl } \mathbf{B}. \tag{70}$$

4.10. The Laplacian operator

The operator

$$\nabla \cdot \nabla = \frac{\partial^2}{\partial x^2} + \frac{\partial^2}{\partial y^2} + \frac{\partial^2}{\partial z^2} \qquad (71)$$

is called the Laplacian operator, and is written ∇^2, read "del squared". It occurs frequently in various branches of physics.

$$\text{div}(\text{grad } \phi) = \nabla \cdot (\nabla \phi) = \nabla^2 \phi.$$

We can also assign a meaning to $\nabla^2 \mathbf{A}$.

$$\begin{aligned}\nabla^2 \mathbf{A} &= \nabla^2 (A_x \mathbf{i} + A_y \mathbf{j} + A_z \mathbf{k}) \\ &= (\nabla^2 A_x)\mathbf{i} + (\nabla^2 A_y)\mathbf{j} + (\nabla^2 A_z)\mathbf{k}. \end{aligned} \qquad (72)$$

For example, when simplifying curl (curl $\mathbf{A}$) we obtain

$$\nabla \times (\nabla \times \mathbf{A}) = \nabla(\nabla \cdot \mathbf{A}) - \nabla^2 \mathbf{A},$$

that is,

$$\text{curl}(\text{curl } \mathbf{A}) = \text{grad div } \mathbf{A} - \nabla^2 \mathbf{A}. \qquad (73)$$

Laplace's equation

$$\nabla^2 \phi = 0 \qquad (74)$$

and Poisson's equation

$$\nabla^2 \phi = -4\pi\rho \qquad (75)$$

are two very important differential equations which occur in several branches of physics.

4.11. Orthogonal curvilinear coordinates

Frequently we have need to use parameters u_1, u_2, u_3 where

$$x = x(u_1, u_2, u_3), \qquad y = y(u_1, u_2, u_3), \qquad z = z(u_1, u_2, u_3)$$

and the surfaces $u_1 = const.$, $u_2 = const.$, $u_3 = const.$ are orthogonal to one another. When we calculate ds^2, the square of the line element ds, we have

$$\begin{aligned} ds^2 &= dx^2 + dy^2 + dz^2 \\ &= h_1^2 \, du_1^2 + h_2^2 \, du_2^2 + h_3^2 \, du_3^2. \end{aligned} \qquad 76)$$

VECTOR CALCULUS

where h_1, h_2, h_3 may be functions of u_1, u_2, u_3. To find grad, div and curl in these coordinates, consider the parallelepiped shown in Fig. 22. Suppose **A** is a vector field with components A_1, A_2, A_3 in the three directions in which u_1, u_2, u_3 increase. Then it may be shown that

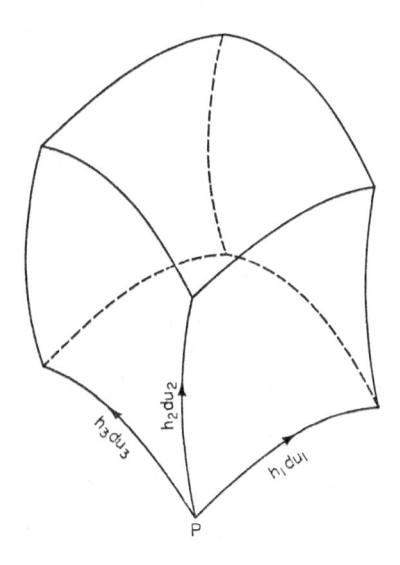

Fig. 22

$$\text{grad } \phi = \left(\frac{1}{h_1}\frac{\partial \phi}{\partial u_1}, \frac{1}{h_2}\frac{\partial \phi}{\partial u_2}, \frac{1}{h_3}\frac{\partial \phi}{\partial u_3}\right) \tag{77}$$

$$\text{div } \mathbf{A} = \frac{1}{h_1 h_2 h_3}\left\{\frac{\partial}{\partial u_1}(h_2 h_3 A_1) + \frac{\partial}{\partial u_2}(h_3 h_1 A_2) + \frac{\partial}{\partial u_3}(h_1 h_2 A_3)\right\} \tag{78}$$

$$(\text{curl } \mathbf{A})_1 = \frac{1}{h_2 h_3}\left\{\frac{\partial}{\partial u_2}(A_3 h_3) - \frac{\partial}{\partial u_3}(A_2 h_2)\right\}, \tag{79}$$

with (curl **A**)$_2$ and (curl **A**)$_2$ obtained by cyclic interchange.

$$\nabla^2 \phi = \frac{1}{h_1 h_2 h_3}$$
$$\times \left[\frac{\partial}{\partial u_1}\left(\frac{h_2 h_3}{h_1} \frac{\partial \phi}{\partial u_1}\right) + \frac{\partial}{\partial u_2}\left(\frac{h_3 h_1}{h_2} \frac{\partial \phi}{\partial u_2}\right) + \frac{\partial}{\partial u_3}\left(\frac{h_1 h_2}{h_3} \frac{\partial \phi}{\partial u_3}\right) \right]. \tag{80}$$

In particular for spherical polar coordinates r, θ, ψ since

$$ds^2 = dr^2 + (r\, d\theta)^2 + (r \sin\theta\, d\psi)^2$$

we take

$$u_1 = r, \qquad u_2 = \theta, \qquad u_3 = \psi$$
$$h_1 = 1, \qquad h_2 = r, \qquad h_3 = r \sin \theta$$

and substitute in (77), (78), (79) and (80) to get grad ϕ, div **A**, curl **A**, and $\nabla^2 \phi$ in spherical polar coordinates.

For cylindrical polar coordinates ρ, ψ, z

$$ds^2 = d\rho^2 + \rho^2\, d\psi^2 + dz^2,$$
$$u_1 = \rho, \qquad u_2 = \psi, \qquad u_3 = 1$$
$$h_1 = 1, \qquad h_2 = \rho, \qquad h_3 = 1.$$

Exercise IV

1. Evaluate $\int_C \mathbf{A}.d\mathbf{s}$ in the following cases:

   (i) $\mathbf{A} = 2xz^2\mathbf{i} + (3y^2 - z)\mathbf{j} + (2zx^2 - y)\mathbf{k}$ and C is the arc of the curve whose parametric equation is

   $$\mathbf{r} = t\mathbf{i} + (t + t^2)\mathbf{j} + (t + t^3)\mathbf{k}$$

   from the point $(0, 0, 0)$ to $(1, 2, 2)$.

   (ii) $\mathbf{A} = z\mathbf{i} + x\mathbf{j} + y\mathbf{k}$ and C is the arc of curve $\mathbf{r} = \cos t\,\mathbf{i} + \sin t\,\mathbf{j} + t\mathbf{k}$ from $t = 0$ to $t = 2\pi$.

   (iii) $\mathbf{A} = (x^3 - y^3)\mathbf{j}$ where C is the arc of the curve $y = 1 - x^2$, $z = 0$ from the point $(1, 0, 0)$ to $(-1, 0, 0)$.

VECTOR CALCULUS

2. Find the circulation of **A** round the curve C where
   (i) $\mathbf{A} = y^3\mathbf{i}$, C is the circle $x^2 + y^2 = a^2$, $z = 0$.
   (ii) $\mathbf{A} = y\mathbf{i} + z\mathbf{j} + x\mathbf{k}$ and C is the circle $x^2 + y^2 = 1$, $z = 0$
   [Cf. Lond. S. 1944]
   (iii) $\mathbf{A} = e^x \sin y\mathbf{i} + e^x \cos y\mathbf{j}$
   C is the rectangle whose vertices are $(0, 0)$, $(1, 0)$, $(1, \pi/2)$, $(0, \pi/2)$.

3. (i) If **a** is a constant vector, and $\mathbf{r} = (x, y, z)$, $r = |\mathbf{r}|$, find grad $(\mathbf{a}.\mathbf{r})$, div $(r^n\mathbf{r})$.
   (ii) Given a field $\phi(\mathbf{r})$, and $\mathbf{A}(\mathbf{r}) = \text{grad } \phi(\mathbf{r})$, show that $\oint \mathbf{A}.d\mathbf{r} = 0$, where the integral is taken along a closed curve.
   (iii) If a vector function $\mathbf{B}(\mathbf{r})$ satisfies

   $$\oint \mathbf{B}.d\mathbf{r} = 0$$

   for every closed curve, then there exists a single valued scalar function of which $B(r)$ is the gradient.

4. **r** is the position vector $(x, y.z)$, **a** is a constant vector, and **v** is defined by $\mathbf{v} = \mathbf{r}(\mathbf{a}.\mathbf{r})$. Show that

   $$\text{div } \mathbf{v} = 4(\mathbf{a}.\mathbf{r}), \text{ curl } \mathbf{v} = \mathbf{a} \times \mathbf{r}.$$

5. State and prove a necessary and sufficient condition for a vector field u to be expressible in the form
   $$u = - \text{ grad } \phi,$$
   where ϕ is a scalar field.
   Determine whether the following vector fields are expressible in this form:

   (i) $(\mathbf{a}.\mathbf{r})\mathbf{r}$, (ii) $(\mathbf{a}.\mathbf{r})\mathbf{a}$,
   (iii) $r^{-1}[r^2\mathbf{a} + (\mathbf{a}.\mathbf{r})\mathbf{r}]$, (iv) $r^{-1}[r^2\mathbf{a} - (\mathbf{a}.\mathbf{r})\mathbf{r}]$,

   where **a** is a constant non-zero vector. Find suitable scalar fields ϕ in appropriate cases. [Camb. P.M. 1956]

6. **F** is the force on a particle placed in a field of force. If the field is conservative, i.e. if the work done on the particle in moving it from a given point to another given point is independent of the path taken, show that curl $\mathbf{F} = \mathbf{O}$. Determine whether this condition is satisfied when

   (i) $\mathbf{F} = \dfrac{\mu\mathbf{r}}{|\mathbf{r}|^3};$ (ii) $\mathbf{F} = \mathbf{k}(\mathbf{k}.\mathbf{r});$

   where **k** is a constant vector and **r** is the radius vector. [Camb. N.S. 1953]

7. Investigate if **A** is a gradient vector in the following cases:
   (i) $\mathbf{A} = (-2z^2x, 3y^2z, y^3 - 2zx^2)$ [Camb. P.N.S. 1947]
   (ii) $\mathbf{A} = \mathbf{k}f(r)$
   (iii) $\mathbf{A} = \mathbf{r}f(r)$. [Camb. N.S. 1957]

8. Distinguish between vector fields which can be derived from scalar potential functions and those which cannot.

Show that the fields

(i) $V = $ constant, (ii) $V = f(r)\hat{\mathbf{r}}$,

are potential fields. Give physical examples of each of these types of fields occurring in (a) Newtonian gravitation, (b) electrostatics, (c) hydrodynamics, describing the fields and giving the potential functions.

If a vector field is given by

$$V = \frac{\hat{\mathbf{k}} \wedge \hat{\mathbf{r}}}{|\hat{\mathbf{k}} \wedge \hat{\mathbf{r}}|^2},$$

the unit vector $\hat{\mathbf{k}}$ being constant, prove that this is not a potential field at the points for which it is defined, but that it is a potential field in the region obtained by excluding all the points of any half-plane bounded by the line $\hat{\mathbf{k}} \wedge \hat{\mathbf{r}} = \mathbf{O}$. [Lond. G.II. 1957]

9. Prove Stokes' theorem

$$\oint \mathbf{F} \cdot d\mathbf{r} = \int \mathbf{n} \cdot \text{curl } \mathbf{F} \, dS,$$

where the symbols have their usual meanings, which should be stated. Exemplify the theorem by calculating the integrals when

$$\mathbf{F} = y\mathbf{i} - x\mathbf{j} + z\mathbf{k}$$

and the surface is the hemisphere $x^2 + y^2 + a^2$ ($z > 0$).

If curl $\mathbf{F}$ vanishes throughout the region between two coaxial infinite circular cylinders, what can be deduced about the circulation of $\mathbf{F}$ around closed curves which lie wholly within this region? [Camb. N.S. 1956]

10. Establish the results:

$$\text{curl } \phi\mathbf{A} = \phi \text{ curl } \mathbf{A} + \text{grad } \phi \times \mathbf{A},$$

$$\text{curl } (\mathbf{A} \times \mathbf{B}) = \mathbf{A} \text{ div } \mathbf{B} - \mathbf{B} \text{ div } \mathbf{A} + (\mathbf{B} \cdot \text{grad})\mathbf{A} - (\mathbf{A} \cdot \text{grad})\mathbf{B}.$$

Evaluate curl $[(\mathbf{a} \times \mathbf{r}/r^n)]$, where $\mathbf{a}$ is a constant vector, $\mathbf{r} = (x, y, z)$ is the position vector of a point, and $r = |\mathbf{r}|$.

11. (i) Prove Green's theorem, namely

$$\iiint \left(\frac{\partial u}{\partial x} + \frac{\partial v}{\partial y} + \frac{\partial w}{\partial z}\right) dx \, dy \, dz = \iint (lu + mv + nw) \, dS$$

where l, m, n are the direction cosines of the outward normal at a point of a closed surface S.

(ii) If S is a sphere of radius a, show that

(i) $\iint lm \, dS = 0$ (ii) $\iint l^2 \, dS = \frac{4}{3}\pi a^2$.

[Cf. Camb. N.S. 1943]

(*iii*) Outline the proof of the theorem

$$\int \text{div } \mathbf{A} \, dV = \int \mathbf{A} \cdot \mathbf{n} \, dS$$

explaining the meaning of the symbols involved.

Evaluate the integral $\iint [x^3 \, dy \, dz + y^3 \, dz \, dx)$ taken over the surface of the sphere $x^2 + y^2 + z^2 = a^2$. [Camb. P.N.S. 1942]

12. $\mathbf{r}(x, y, z)$ is the position vector of a point relative to a given origin 0. V is the volume of the smaller part of the sphere $x^2 + y^2 + z^2 = a^2$, cut off by the plane $z = \tfrac{1}{2}a$. Denoting the bounding surface by S, verify by direct integration of both sides that

$$\int_V \text{div } \mathbf{r} \, dV = \int_S \mathbf{r} \cdot d\mathbf{S}.$$

[Camb. P.N.S. 1945]

13. A sphere whose surface S is of radius a contains a uniform distribution of gravitating matter, and the gravitational field at each point in and on S is $-(g/a)$ times the position-vector of the point with respect to the centre of S, g being a constant. Calculate the flux of the field out of $S(i)$ as a surface-integral over S, and (*ii*) as a volume-integral throughout the interior of S, verifying the equality of the two answers. [Camb. N.S. 1948]

14. State the divergence theorem connecting a surface and a volume integral. Illustrate its application by showing that if ρ and $\mathbf{v}$ are respectively the density and velocity at a point in a moving fluid then div $(\rho \mathbf{v}) + \partial \rho / \partial t = 0$. Verify the equality of the surface and volume integrals for the following vectors, taking the surface to be that of the sphere with centre at the origin and radius a:

 (*i*) $\mathbf{F} = \mu \mathbf{r}$, where $\mathbf{r}$ is the radius vector and μ is a constant.
 (*ii*) $\mathbf{G} = \text{grad } \phi$, where $\phi = x^4 + y^4 + z^4$. [Camb. N.S. 1954]

15. (*i*) Show that div $(\phi \text{ grad } \psi) = \phi \nabla^2 \psi + \text{grad } \phi \cdot \text{grad } \psi$. Hence or otherwise show that if $\nabla^2 \phi = 0$ on and within a simple closed surface S, then $\iint \phi \text{ grad } \phi \cdot \mathbf{n} \, dS \geqslant 0$, where the integral is taken over the surface S, and $\mathbf{n}$ is a unit vector pointing along the outward normal to the surface element dS.

(ii) Show that div $(\mathbf{u} \times \mathbf{v}) = \mathbf{v} \cdot \text{curl } \mathbf{u} - \mathbf{u} \cdot \text{curl } \mathbf{v}$. [Camb. P.N.S. 1947]

16. If $\mathbf{F}$ and f are point-functions, prove that the components of $\mathbf{F}$ normal and tangential to a level surface of f are

$$\frac{(F \cdot \nabla f) \nabla f}{(\nabla f)^2} \quad \text{and} \quad \frac{\nabla f \times (F \times \nabla f)}{(\nabla f)^2}.$$

[Camb. P.M. 1951]

17. A vector function $\mathbf{H}$ of position is defined as follows:

$$\mathbf{H} = \text{curl } \mathbf{A}, \quad \mathbf{A} = \mathbf{k} \log r,$$

where r is the distance from a straight line, and $\mathbf{k}$ is a fixed vector parallel to the line. Prove that, at any point not on the line, div $\mathbf{H} = 0$ and curl $\mathbf{H} = \mathbf{O}$. [Camb. N.S. 1946]

18. (*i*) Prove that
$$\operatorname{curl}\{f(r)\mathbf{r}\} = \mathbf{O}.$$

Find the form of the function $f(r)$ if
$$\operatorname{div}\{f(r)\mathbf{r}\} = 0.$$

(*ii*) Show that, if S is a simple closed surface and if $\mathbf{r}$ is the position vector of any point P on its surface with respect to an origin 0,
$$\int_S r^2 (\mathbf{r}.\mathbf{n})\, dS = \tfrac{5}{2}(A + B + C),$$
where $\mathbf{n}$ is the unit vector along the outward normal at P and A, B, C are the principal moments of inertia at O of a solid of unit density bounded by S. [Lond. G.II. 1959]

19. State, without proof, Stokes' Theorem, and also the divergence theorem (Gauss's Integral Theorem).

If $\mathbf{A} = \operatorname{curl} \mathbf{B}$ and $\mathbf{B} = \operatorname{curl} \mathbf{C}$, and if $\mathbf{C}$ is a suitable vector field, show that
$$\int_v \mathbf{B}^2\, dv = \int_S (\mathbf{C} \times \mathbf{B}).\mathbf{n}\, dS + \int_v \mathbf{C}.\mathbf{A}\, dv,$$
where n is the outward normal to a surface S enclosing a volume v.

If u, $\mathbf{H}$ are suitable scalar and vector fields, prove that
$$\operatorname{curl}(u\mathbf{H}) = \nabla u \times \mathbf{H} + u \operatorname{curl} \mathbf{H}.$$

Hence or otherwise show that
$$\int_\Sigma u\mathbf{n}.\operatorname{curl} \mathbf{H}\, dS = \int_C u\, \mathbf{H}.d\mathbf{s} - \int_\Sigma (\nabla u \times \mathbf{H}).\mathbf{n}\, dS,$$
where $\mathbf{n}$ is the outward normal to a surface Σ which has the closed curve C as boundary. By putting $\mathbf{H} = \nabla w$, or otherwise, show that
$$\int_C u \nabla w.d\mathbf{s} = -\int_C w \nabla u.d\mathbf{s},$$
where w is another scalar field. [Lond. G.II. 1960]

Chapter 5

Some Applications

5.1. Equivalence of force systems

Two systems of forces are said to be *equivalent* if when they separately act upon a rigid body they would produce the same effect in each case.

The equation of motion of a rigid body that is acted upon by forces $F_1, F_2, \ldots, F_n$ at points whose position vectors are $r_1, r_2, \ldots, r_n$, respectively, may be expressed in the form

$$\dot{\mathbf{p}} = \sum \mathbf{F}_i$$
$$\dot{\mathbf{h}} = \sum \mathbf{r}_i \times \mathbf{F}_i$$

where $\mathbf{p}$ is the linear momentum and $\mathbf{h}$ the angular momentum of the body about a fixed origin O.

Hence the conditions of equivalence of two sets of forces
(a) $F_1, F_2, \ldots, F_n$ acting at $r_1, r_2, \ldots, r_n$ and
(b) $F'_1, F'_2, \ldots, F'_m$ acting at $r'_1, r'_2, \ldots, r'_m$ are:

$$\text{(i)} \quad \sum_{i=1}^{n} \mathbf{F}_i = \sum_{j=1}^{m} \mathbf{F}'_j \tag{1}$$

$$\text{(ii)} \quad \sum_{i=1}^{n} \mathbf{r}_i \times \mathbf{F}_i = \sum_{j=1}^{m} \mathbf{r}'_j \times \mathbf{F}'_j. \tag{2}$$

Another way in which the condition of equivalence of two force systems may be expressed is to say that one system, when acting along with the reverse of the other system, upon a rigid body would keep the body in equilibrium.

As a particular case, let us consider

(*a*) a force **F** acting at a point P whose position vector is **r**, and

(*b*) a force **F** acting at a point P' whose position vector is **r**′ where $\overrightarrow{PP'}$ is in the direction of **F**.

Of the conditions (1) and (2) for the equivalence of these two force systems, we see that (1) is automatically satisfied, while (2) gives $\mathbf{r} \times \mathbf{F} = \mathbf{r}' \times \mathbf{F}$, that is

$$(\mathbf{r} - \mathbf{r}') \times \mathbf{F} = \mathbf{O},$$

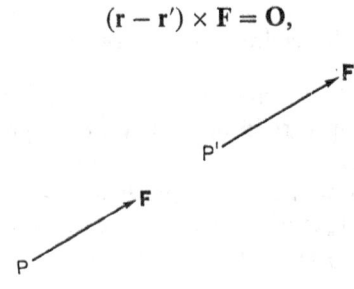

Fig. 1a

which is satisfied since $\mathbf{r} - \mathbf{r}'$ and **F** are in the same direction. Hence we see that **F** at P and **F** at P' are equivalent. This is the principle of transmissibility of a force, namely that a force **F** acting along a line may be regarded as acting at any point on the line.

If two forces $\mathbf{F}_1$ and $\mathbf{F}_2$ act along lines which intersect, they are equivalent to the force $\mathbf{F}_1 + \mathbf{F}_2$ acting at the point of intersection of the two lines. Thus we see that the resultant of two intersecting forces is given by the *Parallelogram of Forces*.

It may be verified that the conditions (1) and (2) for equivalence of systems are satisfied by the systems (*a*) and (*b*) in the following:

(*a*) $\mathbf{F}_1$ at $\mathbf{r}_1$ and $\mathbf{F}_2$ at $\mathbf{r}_2$, where $\mathbf{F}_1$ and $\mathbf{F}_2$ are parallel.

(*b*) Force $\mathbf{F}_1 + \mathbf{F}_2$ (which is parallel to $\mathbf{F}_1$ and $\mathbf{F}_2$) acting at a point

$$\mathbf{r} = \frac{F_1 \mathbf{r}_1 + F_2 \mathbf{r}_2}{F_1 + F_2},$$

where $F_1 + F_2 \neq O$.

SOME APPLICATIONS 105

This shows that the resultant of two parallel forces (which are not equal and opposite) is a third parallel force. The line of action of the resultant forces passes through the centroid of masses proportional to F_1 and F_2 placed at points r_1 and r_2 respectively.

Equal and opposite forces $\mathbf{F}$ and $-\mathbf{F}$ acting along parallel lines constitute a *couple*. A couple has the property that the moment about any point of the pair of forces constituting the couple does not depend on the point about which the moment is taken. If $\mathbf{a}$ and $\mathbf{b}$ are position vectors of the points of application of $\mathbf{F}$ and $-\mathbf{F}$, respectively, the vector moment about any point P whose position vector is $\mathbf{p}$, is

$$\mathbf{G} = (\mathbf{a} - \mathbf{p}) \times \mathbf{F} + (\mathbf{b} - \mathbf{p}) \times (-\mathbf{F})$$
$$= (\mathbf{a} - \mathbf{b}) \times \mathbf{F},$$

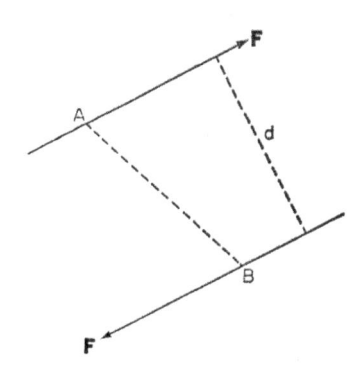

Fig. 1b

showing that $\mathbf{G}$ does not depend on $\mathbf{p}$. The direction of $\mathbf{G}$ is normal to the plane of the forces, and its magnitude is

$$|\mathbf{G}| = F.AB.\sin\theta = Fd.$$

$\mathbf{G}$ is called the moment of the couple.

Two couples of moments $\mathbf{G}$ and $\mathbf{G}'$ one made up of forces $\mathbf{F}$ and $-\mathbf{F}$ acting along one set of parallel lines, and the other couple made up of $\mathbf{F}'$ and $-\mathbf{F}'$ acting along another set of parallel lines, are equivalent if their vector moments are the same, that is, if $\mathbf{G} = \mathbf{G}'$. Since $\mathbf{G}$ and $\mathbf{G}'$ have the same direction both sets of forces lie in the same or in parallel planes. The magnitudes and distance apart of each pair of forces are such that

$$|\mathbf{G}| = Fd = F'd' = |\mathbf{G}'|.$$

It is sufficient to specify a couple by its moment. It is not necessary to indicate the particular pair of forces constituting it.

The sum of two couples of moments $\mathbf{G}$ and $\mathbf{G}'$ is a single couple of moment $\mathbf{G} + \mathbf{G}'$, the moment of the resultant couple being the vector sum of the moments of the two couples.

We now establish the theorem that a force acting along a given line l may be transferred parallel to itself to act at a given point O by introducing a suitable couple.

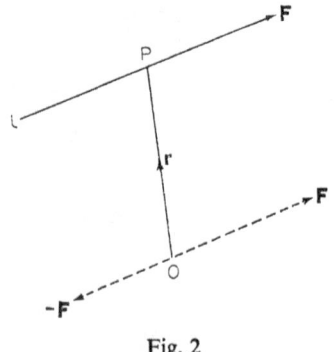

Fig. 2

Introduce at O two equal and opposite forces $\mathbf{F}$ and $-\mathbf{F}$ as shown in Fig. 2. These forces balance each other and do not alter the system. The given force $\mathbf{F}$ along l and the force $-\mathbf{F}$ at O are equal and opposite and act along parallel lines, and constitute a couple $\mathbf{G}$.

The given system is thus equivalent to the force $\mathbf{F}$ at O together with the couple $\mathbf{G}$. It may be noted that the moment $\mathbf{G}$ is the

SOME APPLICATIONS 107

moment about O of the given force **F**, that is, $\mathbf{G} = \mathbf{r} \times \mathbf{F}$ when **r** is the position vector referred to O of a point P on l. By introducing the couple **G**, the force **F** acting along l has been displaced parallel to itself to act at the given point O.

5.2. Poinsot's Central Axis

We consider the general case of a system of forces $\mathbf{F}_1, \mathbf{F}_2, \ldots, \mathbf{F}_n$ acting at points $P_1, P_2, \ldots, P_n$, and seek to reduce the forces to a simple equivalent system.

An arbitrary point is selected as origin O. Let the position vectors of $P_1, \ldots, P_n$ referred to O be $\mathbf{r}_1, \ldots, \mathbf{r}_n$. Each force $\mathbf{F}_j$ may be shifted parallel to itself to act through O by introducing a couple of moment $\mathbf{r}_j \times \mathbf{F}_j, j = 1, \ldots, n$. The forces at O may then be compounded into a single force **R** at O, and the couples may similarly be compounded into a single couple **G**, where

$$\mathbf{R} = \mathbf{F}_1 + \mathbf{F}_2 + \ldots + \mathbf{F}_n$$
$$\mathbf{G} = \mathbf{r}_1 \times \mathbf{F}_1 + \ldots + \mathbf{r}_n \times \mathbf{F}_n.$$

Thus the system is reduced to the force **R** at O and a couple **G**. **R** is the vector sum of the given vectors $\mathbf{F}_1, \ldots, \mathbf{F}_n$ and does not vary with the point O. But the couple **G** depends on the point O chosen, being the sum of the moments about O of the given forces.

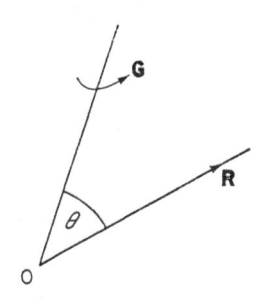

Fig. 3

If it happens that $\mathbf{R} = \mathbf{O}$ and $\mathbf{G} = \mathbf{O}$, then the system of forces is in equilibrium. If $\mathbf{R} = \mathbf{O}$ and $\mathbf{G} \neq \mathbf{O}$, the system is equivalent to the couple $\mathbf{G}$. If $\mathbf{G} = \mathbf{O}$ and $\mathbf{R} \neq \mathbf{O}$, the system is equivalent to the force $\mathbf{R}$ at O.

When $\mathbf{R}$ and $\mathbf{G}$ are non vanishing, they are generally not along the same direction. The angle θ between them is given by

$$\cos \theta = \frac{\mathbf{R}.\mathbf{G}}{RG}.$$

A question for consideration is whether it is possible to simplify this system still further. For example, is it possible to find a point O' such that when the system is reduced to act at O', the resulting couple $\mathbf{G}'$ would vanish? It will be seen that this is generally not possible, except in a particular case which will be considered later. The next possibility of simplification is to find a point O' such that $\mathbf{G}'$ and $\mathbf{R}'$ are in the same direction. If $\overrightarrow{OO'} = \mathbf{r}$, then $\mathbf{R}' = \mathbf{R}$ and $\mathbf{G}'$ which is the moment of the system about O' is the sum of the moment $\mathbf{G}$ and the moment about O' of $\mathbf{R}$ acting at O, and hence

$$\mathbf{G}' = \mathbf{G} + (-\mathbf{r}) \times \mathbf{R}.$$

If O' is such that $\mathbf{G}'$ and $\mathbf{R}$ are in the same direction, then for some scalar p

$$\mathbf{G} - \mathbf{r} \times \mathbf{R} = p\mathbf{R}. \tag{3}$$

Taking scalar product with $\mathbf{R}$,

$$\mathbf{G}.\mathbf{R} = p\mathbf{R}^2$$

Hence $p = (\mathbf{G}.\mathbf{R})/R^2$. The position vector $\mathbf{r}$ of the points O' for which $\mathbf{G}'$ and $\mathbf{R}$ have the same direction satisfy

$$\mathbf{G} - \mathbf{r} \times \mathbf{R} = \frac{\mathbf{G}.\mathbf{R}}{\mathbf{R}^2} \mathbf{R}. \tag{4}$$

These are the equations of a straight line. Taking vector product with **R** we obtain

$$\mathbf{R} \times \mathbf{G} - R^2 \mathbf{r} + (\mathbf{R} \cdot \mathbf{r})\mathbf{R} = \mathbf{O}. \tag{5}$$

We see that

$$\mathbf{r} = \frac{\mathbf{R} \times \mathbf{G}}{R^2} \tag{6}$$

satisfies eqn. (5), and therefore is the position vector of one point on the line. The general solution of (4) is

$$\mathbf{r} = \frac{\mathbf{R} \times \mathbf{G}}{R^2} + \lambda \mathbf{R} \tag{7}$$

where λ is an arbitrary scalar parameter.

Equation (7) gives a line Γ of points O' for which $\mathbf{G}'$ and $\mathbf{R}$ are in the same direction. The direction of this line is the same as that of $\mathbf{R}$. When O' lies on this line, the system of forces is equivalent to a force $\mathbf{R}$ acting along Γ and a couple whose moment is $p\mathbf{R}$ in the same direction. Such a system consisting of a force along a line and a couple about the same line is called a *wrench*. The line Γ is called *Poinsot's Central Axis*. The ratio of the couple to the force is called the *pitch* of the wrench. In the above, the pitch is $p = \mathbf{G} \cdot \mathbf{R}/R^2$. In scalars, if $\mathbf{r} = (x, y, z)$, $\mathbf{R} = (X, Y, Z)$, $\mathbf{G} = (L, M, N)$ the equations of the central axis are

$$\frac{L - yZ + zY}{X} = \frac{M - zX + xZ}{Y} = \frac{N - xY + yX}{Z}$$
$$= \frac{LX + MY + NZ}{X^2 + Y^2 + Z^2} = p. \tag{8}$$

In Fig. 4, the couple G is resolved into $G \cos \theta$ and $G \sin \theta$. The effect of the couple $G \sin \theta$ is to displace the force $\mathbf{R}$ at O parallel to itself through a distance $G \sin \theta / R$, thus giving the axis of the wrench. The couple $G \cos \theta$ is equivalent to couple $p\mathbf{R}$ about the axis of the wrench.

Invariants

The vector $\mathbf{R}$ does not depend on the choice of O. Further, the magnitude of $\mathbf{R}$ does not depend on choice of the particular

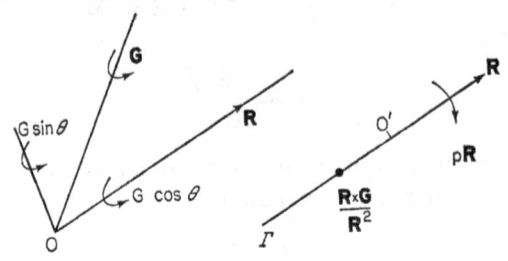

Fig. 4

directions of the axes of coordinates Ox, Oy, Oz. Hence $\mathbf{R}^2 = X^2 + Y^2 + Z^2$ is an invariant with respect to the axes.

Again, we note that
$$\mathbf{G'}.\mathbf{R'} = \mathbf{G}.\mathbf{R},$$
and hence $\mathbf{G}.\mathbf{R} = LX + MY + NZ$ is also an invariant of the system. These two invariants
$$\mathbf{R}^2 = X^2 + Y^2 + Z^2, \qquad \mathbf{R}.\mathbf{G} = LX + MY + NZ$$
are useful in discussing some of the properties of the system.

In the case when $\mathbf{G}$ and $\mathbf{R}$ are such that $\mathbf{G}.\mathbf{R} = O$, then it is possible to choose O' so that $\mathbf{G'} = \mathbf{O}$, and the system is then equivalent to a single force. $\mathbf{R}$ and $\mathbf{G}$ are perpendicular in this case, and the effect of $\mathbf{G}$ is merely to shift the force $\mathbf{R}$ parallel to itself. The equation of the line of action is then
$$\mathbf{G} - \mathbf{r} \times \mathbf{R} = \mathbf{O}$$
which may also be expressed in the form (7).

A system of coplanar forces satisfies this property $\mathbf{G}.\mathbf{R} = O$. If we take axes Ox, Oy in the plane of the forces, and Oz normal to the plane,
$$\mathbf{R} = (X, Y, O), \qquad \mathbf{G} = (O, O, G)$$
with $\mathbf{r} = (x, y, O)$ the equation of the line of action of the resultant is
$$G - xY + yX = O.$$

SOME APPLICATIONS

Example 1. *Two forces P, Q act along the straight lines* $y = x\tan\alpha$, $z = c$; $y = -x\tan\alpha$, $z = -c$ *respectively. Find the equations of the central axis and the pitch of the equivalent wrench.*

Show that as P, Q vary in magnitude the central axis lies on the surface

$$z(x^2 + y^2) = 2cxy \operatorname{cosec} 2\alpha.$$

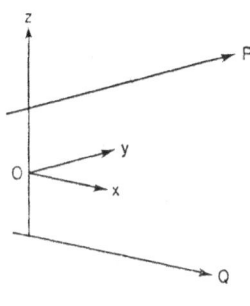

Fig. 5

The equations of the lines may be written in the form

$$\frac{x}{\cos\alpha} = \frac{y}{\sin\alpha} = \frac{z-c}{O}; \qquad \frac{x}{\cos\alpha} = \frac{y}{-\sin\alpha} = \frac{z+c}{O}.$$

$$(X, Y, Z) = \sum \mathbf{F}_i, \qquad (L, M, N) = \mathbf{G} = \sum \mathbf{r}_i \times \mathbf{F}_i$$

Hence

$$X = (P+Q)\cos\alpha, \qquad Y = (P-Q)\sin\alpha, \qquad Z = O$$

$$L = -cP\sin\alpha - cQ\sin\alpha, \quad M = cP\cos\alpha - cQ\cos\alpha, \quad N = O.$$

$$p = \frac{LX + MY + NZ}{X^2 + Y^2 + Z^2}$$

$$= \frac{-c\sin\alpha\cos\alpha(P+Q)^2 + c\sin\alpha\cos\alpha(P-Q)^2}{(P+Q)^2\cos^2\alpha + (P-Q)^2\sin^2\alpha}$$

$$= \frac{-2cPQ\sin 2\alpha}{P^2 + Q^2 + 2PQ\cos 2\alpha}.$$

The equations of the central axis are

$$\frac{-c(P+Q)\sin\alpha - z(P-Q)\sin\alpha}{(P+Q)\cos\alpha} = \frac{c(P-Q)\cos\alpha - z(P+Q)\cos\alpha}{(P-Q)\sin\alpha}$$
$$= \frac{-x(P-Q)\sin\alpha + y(P+Q)\cos\alpha}{0}.$$

These reduce to

$$z\left(\frac{P-Q}{P+Q}\tan\alpha + \frac{P+Q}{P-Q}\cot\alpha\right) = c(\tan\alpha + \cot\alpha)$$

$$y = \frac{P-Q}{P+Q} x \tan\alpha.$$

Eliminating $(P-Q)/(P+Q)$, we obtain

$$z\left(\frac{y}{x} + \frac{x}{y}\right) = c(\tan\alpha + \cot\alpha)$$

that is,

$$(x^2 + y^2)z = 2cxy \operatorname{cosec} 2\alpha.$$

Hence the central axis lies on this surface.

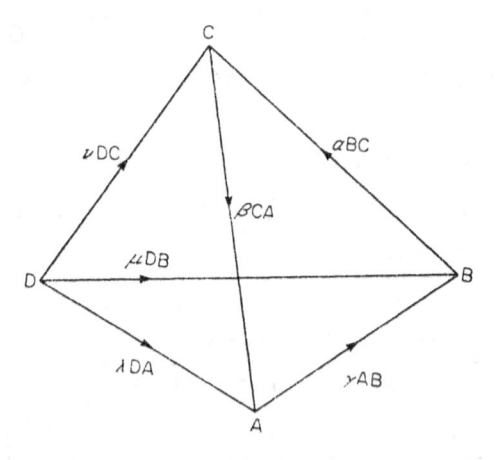

Fig. 6

SOME APPLICATIONS 113

Example 2. *Forces* $\alpha\overrightarrow{BC}$, $\beta\overrightarrow{CA}$, $\gamma\overrightarrow{AB}$, $\lambda\overrightarrow{DA}$, $\mu\overrightarrow{DB}$, $\nu\overrightarrow{DC}$ *act along the edges of a tetrahedron ABCD. If they reduce to a single force show that*

$$\lambda\alpha + \mu\beta + \nu\gamma = O.$$

Suppose D is taken as origin, and the position vectors of A, B, C are **a**, **b**, **c** respectively. The forces then are

$$\alpha(\mathbf{c} - \mathbf{b}), \beta(\mathbf{a} - \mathbf{c}), \gamma(\mathbf{b} - \mathbf{a}), \lambda\mathbf{a}, \mu\mathbf{b}, \nu\mathbf{c}.$$

Hence

$$\mathbf{R} = \sum \mathbf{F} = \alpha(\mathbf{c} - \mathbf{b}) + \beta(\mathbf{a} - \mathbf{c}) + \gamma(\mathbf{b} - \mathbf{a}) + \lambda\mathbf{a} + \mu\mathbf{b} + \nu\mathbf{c}$$

$$= (\beta - \gamma + \lambda)\mathbf{a} + (\gamma - \alpha + \mu)\mathbf{b} + (\alpha - \beta + \nu)\mathbf{c}$$

$$\mathbf{G} = \sum \mathbf{r} \times \mathbf{F}$$

$$= \mathbf{b} \times \alpha(\mathbf{c} - \mathbf{b}) + \mathbf{c} \times \beta(\mathbf{a} - \mathbf{c}) + \mathbf{a} \times \gamma(\mathbf{b} - \mathbf{a})$$

$$= \alpha\mathbf{b} \times \mathbf{c} + \beta\mathbf{c} \times \mathbf{a} + \gamma\mathbf{a} \times \mathbf{b}.$$

The condition that the system may reduce to a single force is $\mathbf{G} \cdot \mathbf{R} = 0$, that is

$$[(\beta - \gamma + \lambda)\alpha + (\gamma - \alpha + \mu)\beta + (\alpha - \beta + \nu)\gamma][\mathbf{a}, \mathbf{b}, \mathbf{c}] = 0.$$

Since **a**, **b**, **c** are not coplanar, $[\mathbf{a}, \mathbf{b}, \mathbf{c}] \neq 0$. The condition reduces to

$$\lambda\alpha + \mu\beta + \nu\gamma = 0.$$

Dynamics
5.3. Space-curve

Suppose the point P whose position vector **r** lies on a twisted curve C in space, and the arc length $P_0 P$ is s, where P_0 is a fixed point on C.

The tangent at P to the curve is defined as the limit as P_1 tends to P of the chord PP_1 where P_1 is a point close to P. Denoting by

Fig. 7

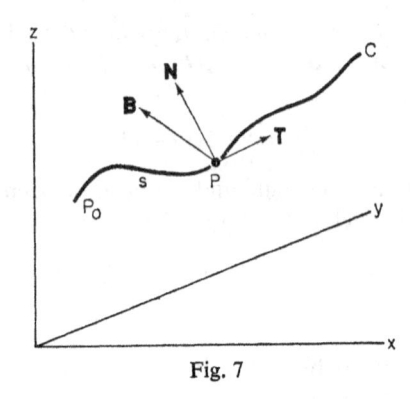

dashes differentiation with respect to s, the unit vector along the tangent at P has been shown in Chapter III to be

$$\mathbf{T} = \frac{d\mathbf{r}}{ds} = \mathbf{r}'.$$

The osculating plane at P is defined by considering a plane PP_1P_2 where P_1, P_2 are points on C close to P, and taking the limit as P_1 and P_2 tend to P. If $\mathbf{R}$ is the position vector of a point on this plane,

$$[\mathbf{R} - \mathbf{r}, \mathbf{r}', \mathbf{r}''] = 0 \quad \text{or} \quad [\mathbf{R} - \mathbf{r}, \mathbf{T}, \mathbf{T}'] = 0. \tag{9}$$

The osculating plane contains the tangent vector $\mathbf{T}$ and its derivative vector $\mathbf{T}'$. Of the normals to the curve at P, there is one which lies in the osculating plane. This is called the *principal normal*, and unit vector along it is denoted by $\mathbf{N}$. Differentiating $\mathbf{T}^2 = 1$ gives $\mathbf{T} \cdot \mathbf{T}' = 0$, showing that $\mathbf{T}'$ is perpendicular to $\mathbf{T}$ and is thus along a normal at P. Also we have seen that $\mathbf{T}'$ lies in the osculating plane. Hence $\mathbf{T}'$ is along the principal normal at P. We write

$$\mathbf{T}' = \kappa \mathbf{N} \tag{10}$$

where the scalar $\kappa = |\mathbf{T}'|$. The normal at P to the osculating plane is called the *binormal* at P, and unit vector along it is denoted by $\mathbf{B}$. $\mathbf{B}$ has the same direction as $\mathbf{T} \times \mathbf{T}'$.

T, **N**, **B** form an orthogonal triad of vectors, and their positive directions taken so that **T**, **N**, **B**, in that order form a positive triad. As P moves along the curve, the triad twists in direction.

The rates at which these vectors change with s are seen to satisfy the equations

$$\mathbf{T}' = \kappa\mathbf{N}, \mathbf{N}' = \tau\mathbf{B} - \kappa\mathbf{T}, \mathbf{B}' = -\tau\mathbf{N}, \tag{11}$$

which are called *Serret–Frenet formulae*. κ is the rate of rotation of the triad about **B** and τ is the rate of rotation about **T**. κ is called the curvature and τ the torsion. $\rho = 1/\kappa$ is called the radius of curvature and $\sigma = 1/\tau$ the radius of torsion.

For a plane curve, the osculating plane is the plane of the curve, the binomal is normal to this plane, the torsion vanishes and ρ is the radius of curvature.

It is useful to obtain expressions for the velocity and acceleration of the point P as it moves along the curve C. The velocity of P is

$$\mathbf{v} = \frac{d\mathbf{r}}{dt} = \frac{ds}{dt}\frac{d\mathbf{r}}{ds} = v\mathbf{T},$$

the acceleration of P is

$$\mathbf{f} = \frac{d\mathbf{v}}{dt} = \frac{d}{dt}(v\mathbf{T}) = \frac{dv}{dt}\mathbf{T} + v^2\frac{d\mathbf{T}}{ds}$$

$$= \frac{dv}{dt}\mathbf{T} + \frac{v^2}{\rho}\mathbf{N}.$$

Thus the components of velocity and acceleration along the directions of the triad are

$$\mathbf{v} = (v, 0, 0)$$

$$\mathbf{f} = \left(\frac{dv}{dt}, \frac{v^2}{\rho}, 0\right).$$

The velocity is along the tangent, and the acceleration is in the osculating plane.

Example 1. *A space-curve is given by the parametric equations*
$$x = a \cos \theta, \qquad y = a \sin \theta, \qquad z = c\theta.$$
Find the unit vectors **T**, **N**, **B**, *the curvature κ and torsion τ.*

$$\mathbf{r} = a \cos \theta \mathbf{i} + a \sin \theta \mathbf{j} + c\theta \mathbf{k}$$

$$\frac{d\mathbf{r}}{d\theta} = -a \sin \theta \mathbf{i} + a \cos \theta \mathbf{j} + c\mathbf{k}$$

$$\frac{ds}{d\theta} = \left|\frac{d\mathbf{r}}{d\theta}\right| = \sqrt{(a^2 \sin^2 \theta + a^2 \cos^2 \theta + c^2)} = \sqrt{(a^2 + c^2)}$$

$$\mathbf{T} = \frac{d\mathbf{r}}{ds} = \frac{1}{\sqrt{(a^2 + c^2)}}(-a \sin \theta \mathbf{i} + a \cos \theta \mathbf{j} + c\mathbf{k}).$$

$$\frac{d\mathbf{T}}{ds} = \frac{1}{(a^2 + c^2)}(-a \cos \theta \mathbf{i} - a \sin \theta \mathbf{j}) = \kappa \mathbf{N}.$$

Hence
$$\kappa = \left|\frac{d\mathbf{T}}{ds}\right| = \frac{a}{a^2 + c^2}, \qquad \mathbf{N} = (-\cos \theta \mathbf{i} - \sin \theta \mathbf{j}).$$

$$\mathbf{B} = \mathbf{T} \times \mathbf{N} = \frac{1}{\sqrt{(a^2 + c^2)}}(c \sin \theta \mathbf{i} - c \cos \theta \mathbf{j} + a\mathbf{k}).$$

$$\tau \mathbf{N} = -\frac{d\mathbf{B}}{ds} = \frac{-1}{a^2 + c^2}(c \cos \theta \mathbf{i} + c \sin \theta \mathbf{j}).$$

$$= +\frac{c}{a^2 + c^2} \mathbf{N}.$$

Thus $\tau = c/(a^2 + c^2)$.

Example 2. *Show that for any space-curve* $[\mathbf{r}', \mathbf{r}'', \mathbf{r}'''] = \kappa^2 \tau.$

$$\mathbf{T}' = \kappa \mathbf{N}, \qquad \mathbf{T}'' = \kappa \mathbf{N}' + \kappa' \mathbf{N} = \kappa(\tau \mathbf{B} - \kappa \mathbf{T}) + \kappa' \mathbf{N}$$

$$\text{L.H.S.} = [\mathbf{T}, \mathbf{T}', \mathbf{T}''] = [\mathbf{T}, \kappa \mathbf{N}, -\kappa^2 \mathbf{T} + \kappa' \mathbf{N} + \kappa \tau \mathbf{B}]$$

$$= \kappa^2 \tau [\mathbf{T}, \mathbf{N}, \mathbf{B}] = \kappa^2 \tau.$$

SOME APPLICATIONS 117

Example 3. *If a particle moving along a space-curve has velocity* **v** *and acceleration* **f**, *show that the radius of curvature* ρ *is*

$$\frac{v^3}{|\mathbf{v} \times \mathbf{f}|}.$$

$$\mathbf{v} = \frac{dv}{dt}\mathbf{T}$$

$$\mathbf{f} = \dot{v}\mathbf{T} + \frac{v^2}{\rho}\mathbf{N}.$$

Hence

$$\mathbf{v} \times \mathbf{f} = \frac{v^3}{\rho}\mathbf{T} \times \mathbf{N} = \frac{v^3}{\rho}\mathbf{B}.$$

Therefore

$$|\mathbf{v} \times \mathbf{f}| = \frac{v^3}{\rho}.$$

5.4. Infinitesimal rotations. Angular velocity

Consider the rotation of a point P about an axis OA through an angle $\delta\phi$. Suppose **e** is unit vector along OA. If PN is the perpendicular from P on OA, the rotation moves P to P_1 along the

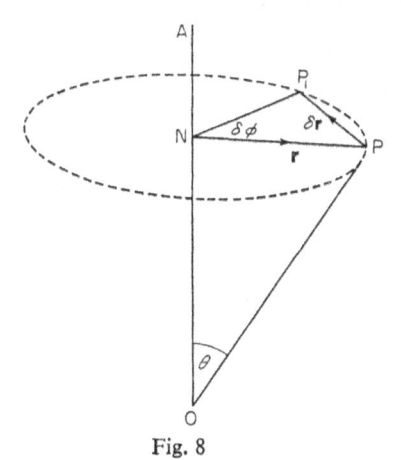

Fig. 8

arc of a circle which has centre N, radius NP, and where OA is normal to the plane of the circle, and $\widehat{PNP_1} = \delta\phi$.
Suppose the position vector of P is $\mathbf{r}$, and of P_1 is $\mathbf{r} + \delta\mathbf{r}$.

$$|\delta\mathbf{r}| = PP_1 = 2PN \sin(\tfrac{1}{2}\delta\phi).$$

For small $\delta\phi$,

$|\delta\mathbf{r}|$ is approximately

$$2r \sin\theta(\tfrac{1}{2}\delta\phi) = r \sin\theta\,\delta\phi.$$

Also the direction of $\delta\mathbf{r}$ is perpendicular to OA, and also perpendicular to OP in the limit $\delta\phi \to 0$.

Hence for small $\delta\phi$

$$\delta\mathbf{r} = \mathbf{e} \times \mathbf{r}\,\delta\phi \tag{12}$$

and

$$\mathbf{r}_1 = \mathbf{r} + (\mathbf{e}\delta\phi) \times \mathbf{r}. \tag{13}$$

If we consider two successive infinitesimal rotations $\delta\phi_1$ about $\mathbf{e}_1$, taking P to P_1, followed by rotation $\delta\phi_2$ about $\mathbf{e}_2$ taking P_1 to P_2, we can show that the order of the two rotations can be interchanged. (This result is not valid for finite rotations.)

$$\mathbf{r}_1 = \mathbf{r} + (\mathbf{e}\delta\phi_1) \times \mathbf{r}$$
$$\mathbf{r}_{12} = \mathbf{r}_1 + (\mathbf{e}_2\delta\phi_2) \times \mathbf{r}_1$$
$$= \mathbf{r} + (\mathbf{e}_1\delta\phi_1) \times \mathbf{r} + \mathbf{e}_2\delta\phi_2 \times \{\mathbf{r} + (\mathbf{e}_1\delta\phi_1) \times \mathbf{r}\}$$
$$= \mathbf{r} + (\mathbf{e}_1\delta\phi_1 + \mathbf{e}_2\delta\phi_2) \times \mathbf{r} \tag{14}$$

neglecting terms in $\delta\phi_1\delta\phi_2$. Thus

$$\mathbf{r}_{12} = \mathbf{r}_{21}. \tag{14A}$$

By considering a rotation $\delta\phi$ about axis $\mathbf{e}$ taking place in time δt, and proceeding to the limit δt tending to zero, thus defining $d\phi/dt$, we obtain the vector

$$\boldsymbol{\omega} = \dot{\phi}\mathbf{e} \tag{15}$$

which is called the angular velocity vector.

SOME APPLICATIONS

Equation (12) then gives

$$\delta\mathbf{r} = \boldsymbol{\omega} \times \mathbf{r}\delta t.$$

Thus the velocity of the point P is

$$\mathbf{v} = \boldsymbol{\omega} \times \mathbf{r}. \tag{16}$$

The magnitude of this velocity is $|\boldsymbol{\omega}||\mathbf{r}|\sin\theta = \omega.NP$. The result (16) being applicable to all points on the line OP, we may speak of $\boldsymbol{\omega}$ as the angular velocity vector of the line OP.

From eqn. (14) we also deduce that angular velocities are commutative, and that the velocity of a point P due to two angular velocities ω_1 and ω_2 is

$$\dot{\mathbf{r}} = \lim_{\delta t \to 0} \frac{\mathbf{r}_{12} - \mathbf{r}}{\delta t} = (\dot{\phi}_1 \mathbf{e}_1 + \dot{\phi}_2 \mathbf{e}_2) \times \mathbf{r}$$
$$= (\boldsymbol{\omega}_1 + \boldsymbol{\omega}_2) \times \mathbf{r},$$

showing that the velocity is the same as that due to the single angular velocity $\omega_1 + \omega_2$. Thus angular velocities about the same point may be compounded or resolved into components according to the usual procedure for vectors.

5.5. Angular velocity of a rigid body

If a rigid body is free to rotate about a fixed point O it may be shown that there exists an angular velocity vector $\boldsymbol{\omega}$ along a line through O such that the velocity of any point P of the body may be regarded as due to the angular velocity $\boldsymbol{\omega}$; that is, the velocity $\dot{\mathbf{r}}$ of P is given by

$$\dot{\mathbf{r}} = \boldsymbol{\omega} \times \mathbf{r}. \tag{17}$$

Suppose that such an ω exists for two points $\mathbf{r}_1$ and $\mathbf{r}_2$ of the body, with

$$\dot{\mathbf{r}}_1 = \boldsymbol{\omega} \times \mathbf{r}_1, \qquad \dot{\mathbf{r}}_2 = \boldsymbol{\omega} \times \mathbf{r}_2.$$

Since the points O, P_1, P_2 are on a rigid body,

$$\mathbf{r}_1^2, \mathbf{r}_2^2 \quad \text{and} \quad (\mathbf{r}_1 - \mathbf{r}_2)^2$$

are all constants. Hence differentiating with respect to t,

$$\mathbf{r}_1 \cdot \dot{\mathbf{r}}_1 = \mathbf{r}_2 \cdot \dot{\mathbf{r}}_2 = (\mathbf{r}_1 - \mathbf{r}_2) \cdot (\dot{\mathbf{r}}_1 - \dot{\mathbf{r}}_2) = 0$$

from which we also obtain

$$\mathbf{r}_1 \cdot \dot{\mathbf{r}}_2 + \mathbf{r}_2 \cdot \dot{\mathbf{r}}_1 = 0.$$

ω is perpendicular to both $\dot{\mathbf{r}}_1$ and $\dot{\mathbf{r}}_2$, and hence we may write

$$\omega = \alpha(\dot{\mathbf{r}}_1 \times \dot{\mathbf{r}}_2)$$

for some α. Substituting for ω

$$\dot{\mathbf{r}}_1 = -\alpha \mathbf{r}_1 \times (\dot{\mathbf{r}}_1 \times \dot{\mathbf{r}}_2) = -\alpha(\mathbf{r}_1 \cdot \dot{\mathbf{r}}_2)\dot{\mathbf{r}}_1$$
$$\dot{\mathbf{r}}_2 = \alpha(\dot{\mathbf{r}}_1 \cdot \mathbf{r}_2)\dot{\mathbf{r}}_2.$$

Thus

$$\alpha = 1/(\dot{\mathbf{r}}_1 \cdot \mathbf{r}_2) \quad \text{and}$$

$$\omega = \frac{\dot{\mathbf{r}}_1 \times \dot{\mathbf{r}}_2}{\dot{\mathbf{r}}_1 \cdot \mathbf{r}_2} = \frac{\dot{\mathbf{r}}_2 \times \dot{\mathbf{r}}_1}{\dot{\mathbf{r}}_2 \cdot \mathbf{r}_1}.$$

Let $\mathbf{r}$ be a third point of the rigid body. Then conditions of rigidity are

$$\mathbf{r} \cdot \dot{\mathbf{r}} = \mathbf{r} \cdot \dot{\mathbf{r}}_1 + \dot{\mathbf{r}} \cdot \mathbf{r}_1 = \mathbf{r} \cdot \dot{\mathbf{r}}_2 + \dot{\mathbf{r}} \cdot \mathbf{r}_2 = 0.$$

Writing $\omega \times \mathbf{r} = \mathbf{q}$, it is easy to deduce that

$$(\mathbf{q} - \dot{\mathbf{r}}) \cdot \mathbf{r} = 0, \quad (\mathbf{q} - \dot{\mathbf{r}}) \cdot \mathbf{r}_1 = 0, \quad (\mathbf{q} - \dot{\mathbf{r}}) \cdot \mathbf{r}_2 = 0.$$

If $\mathbf{r}, \mathbf{r}_1, \mathbf{r}_2$ are not coplanar, $\mathbf{q} - \dot{\mathbf{r}}$ cannot simultaneously be perpendicular to all of them, and hence it is the null vector. That is,

$$\dot{\mathbf{r}} = \mathbf{q} = \omega \times \mathbf{r}$$

for all $\mathbf{r}$.

If $\mathbf{r}, \mathbf{r}_1, \mathbf{r}_2$ are coplanar, we may write

$$\mathbf{r} = \lambda \mathbf{r}_1 + \mu \mathbf{r}_2$$

where λ, μ are constant scalars for given $\mathbf{r}, \mathbf{r}_1, \mathbf{r}_2$, hence

$$\dot{\mathbf{r}} = \lambda \dot{\mathbf{r}}_1 + \mu \dot{\mathbf{r}}_2 = \lambda \omega \times \mathbf{r}_1 + \mu \omega \times \mathbf{r}_2$$
$$= \omega \times (\lambda \mathbf{r}_1 + \mu \mathbf{r}_2) = \omega \times \mathbf{r}.$$

SOME APPLICATIONS 121

Hence a vector ω exists so that the motion of all points of the body may be regarded as due to the angular velocity ω about O.

The line through the origin in the direction of ω is such that every point on the line has no velocity. This line is called the *axis of rotation*. When ω is not constant in time, the direction of ω is called the *instantaneous axis*.

In the more general case when there is no point of the body fixed, the motion may be specified by the velocity $\mathbf{u}$ of a point O of the body together with an angular velocity ω of the body about the point O. The velocity of a point P of the body where $\overrightarrow{OP} = \mathbf{r}$ is

$$\mathbf{v} = \mathbf{u} + \omega \times \mathbf{r}. \tag{18}$$

It can be shown that if a different point O' of the body is used as origin, and $\mathbf{u}'$, ω', the corresponding linear and angular velocities, then $\omega' = \omega$, $\mathbf{u}' = \mathbf{u} + \omega \times \mathbf{a}$. The effect of a rotation ω about an axis through O is equivalent to that of a rotation ω about a parallel axis through O' together with a translation $\omega \times \overrightarrow{OO'}$.

It is possible by choosing O' suitably to reduce the motion of the body to one consisting of a translation along a certain line and a rotation about the same line. Such a motion is called a *screw* motion, the line is called the axis of the screw, and the pitch of the screw is

$$p = |\mathbf{u}'|/|\omega| = \frac{\mathbf{u} \cdot \omega}{\omega^2}. \tag{19}$$

The equation of the axis may be written in the form

$$\mathbf{u} + \omega \times \mathbf{r} = p\omega \tag{20}$$

or alternatively as

$$\mathbf{r} = \frac{\omega \times \mathbf{u}}{\omega^2} + \lambda\omega \tag{21}$$

where λ is a parameter.

When p vanishes the motion is one of pure rotation about the axis.

There is a close analogy between the mathematical relations which occur in rigid-body motion in kinematics, and those concerning forces in statics. The angular velocity ω of the body is analogous to the vector sum **R** of a system of forces, and the velocity of a point $\mathbf{r}_0$ of the body is analogous to the moment about $\mathbf{r}_0$ of the system of forces.

For a plane lamina rotating in its own plane, the angular velocity vector ω is along the normal to the plane. If **n** is unit vector along the normal, and ϕ is the angle which a line fixed in the body makes with a line fixed in space in the plane of the lamina, then

$$\omega = \dot{\phi}\mathbf{n}.$$

The point I where the instantaneous axis meets the plane is called the *instantaneous centre*. The velocity of a point P on the lamina is $\omega \times \overrightarrow{IP}$, which is along the plane and perpendicular to IP and of magnitude $IP \cdot \dot{\phi}$.

Example 1. *Two concentric spheres of radii a, b ($b > a$) are made to rotate with constant angular velocities ω_1, ω_2, respectively, about fixed diameters. A sphere of centre C and radius $\frac{1}{2}(b - a)$ rolls in contact with both spheres. Show that the centre C describes a circle with constant angular velocity.* [*cf.* Lond. S. 1946].

Let **e** be unit vector along OC, and Ω the angular velocity of the rolling sphere. The position vector of C is $\frac{1}{2}(a + b)\mathbf{e}$, and hence the velocity of C is

$$\mathbf{u} = \tfrac{1}{2}(a + b)\dot{\mathbf{e}} \tag{22}$$

The condition for rolling is that the points of contact of the two bodies have the same velocity. Hence the condition for rolling at A is

$$\mathbf{u} + \Omega \times \tfrac{1}{2}(b - a)\mathbf{e} = \omega_2 \times b\mathbf{e} \tag{23}$$

and at B

$$\mathbf{u} - \Omega \times \tfrac{1}{2}(b - a)\mathbf{e} = \omega_1 \times a\mathbf{e}. \tag{24}$$

Adding (23) and (24)

$$2\mathbf{u} = (a\omega_1 + b\omega_2) \times \mathbf{e}$$

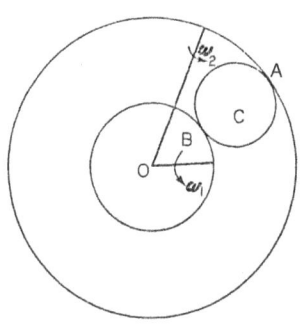

Fig. 9

which may be written

$$\mathbf{u} = \frac{a\omega_1 + b\omega_2}{a + b} \times \tfrac{1}{2}(a + b)\mathbf{e},$$

showing that the motion of the point C may be regarded as due to the angular velocity

$$\omega = \frac{a\omega_1 + b\omega_2}{a + b}.$$

ω is fixed in direction and magnitude, and hence the point C describes a circle with constant speed.

Example 2. *A right circular cone of vertex O and semi-vertical angle α is rolling on a horizontal plane so that at any instant it touches the plane along a generator. If $\mathbf{n}$ is the angular velocity of the cone about its axis, ω the angular velocity of the cone about the generator in contact and $\mathbf{\Omega}$ the angular velocity of the vertical plane through the axis of the cone, show that*

$$n = \omega \cos \alpha, \qquad \Omega = -\omega \tan \alpha.$$

The generator OA is the instantaneous axis. If v is the velocity of C, calculating it in different ways, we obtain

$$v = \Omega . NC = -\omega MC$$

Hence
$$\Omega = -\omega MC/NC = -\omega \tan \alpha.$$

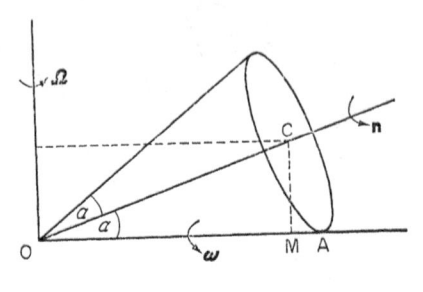

Fig. 10

Since the point A is instantaneously at rest, and velocity of A is the velocity of A relative to C plus the velocity of C, we obtain

$$0 = CA \cdot n + v$$
$$n = -v/CA = \omega \cdot CM/CA = \omega \cos \alpha.$$

Theory of Potential
5.6. Gauss's theorem

In many branches of Applied Mathematics we are concerned with the study of fields, such as electric field, magnetic field, gravitational field and so on. These fields are produced by sources (and sinks).

Thus in electrostatics, the field is produced by electric charges, and in gravitation by masses of matter, in hydromechanics by sources or sinks of fluid. The sources or sinks may be distributed through a volume or on a surface or on a curved line, and we therefore speak of volume distributions, surface distributions and line distributions. For a volume distribution we define a volume density ρ at a point P so that the source in a small volume $d\Omega$ surrounding P is $\rho \, d\Omega$. Though matter is granular in structure, it is convenient to replace the actual matter by an equivalent distribution with a continuous density function, such that the two distributions may have the same average properties. Similarly, the surface density is defined by a continuous function σ on the surface such that the source in area dS is $\sigma\, dS$, and line density λ such that source in length ds is $\lambda\, ds$. It is also frequently useful to consider sources as occupying geometrical points, in which case they are called point-sources, and their strength is denoted by symbols such as e or m.

The relationships between the sources and the fields produced by them have been derived from experiment, and have been formulated in two alternative ways. One uses the result that the force between two sources is given by the *Inverse Square Law*. The other uses the property that the flux of the field is enunciated by *Gauss's theorem*. The two formulations are mathematically equivalent.

Taking for discussion the gravitational field produced by masses, the Inverse Square Law of gravitation states that the force between two masses m_1 and m_2 at a distance r apart is an attractive force of amount $\gamma m_1 m_2/r^2$. The gravitational field at any point is defined as the force a unit mass would experience when placed at that point. The field **F** at a point P, whose position vector is **r**, due to a point-mass m placed at the origin 0, is

$$\mathbf{F} = -\gamma \frac{m\mathbf{r}}{r^3}. \tag{25}$$

For a system of point-masses m_1 at P_1, m_2 at P_2, ... m_n at P_n, the field at P is

$$\mathbf{F} = -\gamma \sum_{i=1}^{n} \frac{m_i(\mathbf{r} - \mathbf{r}_i)}{|\mathbf{r} - \mathbf{r}_i|^3}. \tag{26}$$

Gauss's theorem states that the outward flux of the field vector across any closed surface is

$$N = -4\pi\gamma M_i, \tag{27}$$

where M_i is the total mass enclosed by the surface, that is,

$$\int_S \mathbf{F}.d\mathbf{S} = -4\pi\gamma \quad \text{(mass enclosed)}.$$

Gauss's theorem may be derived from the Inverse Square Law and vice versa.

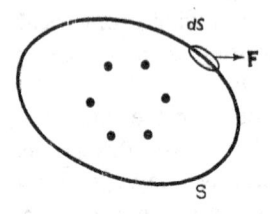

Fig. 11

Assuming Gauss's theorem, to derive Inverse Square Law, consider the field due to a particle m at O, and apply Gauss's theorem to a sphere of centre O and radius r (Fig. 12). By symmetry, $\mathbf{F}$ is radial, and its magnitude is the same at all points of the sphere. Hence the flux is $N = 4\pi r^2 F$. Hence Gauss's theorem gives

$$4\pi r^2 . F = -4\pi\gamma . m,$$

$$F = -\frac{\gamma m}{r^2}.$$

Hence

$$\mathbf{F} = -\frac{\gamma m \mathbf{r}}{r^3}.$$

Fig. 12

To prove the converse result, it is helpful to use certain properties of solid angles. Suppose S is a simple open surface, so that no straight line meets S in more than two points, and let the closed curve C be the boundary of S.

The lines joining a given point A to points of C generate a cone. The area cut off by this cone on a sphere of centre A and unit radius is called the *solid angle* ω subtended by the surface S at the point A. If dS is surface element at P, ω the solid angle subtended by dS at A, and θ the angle between the radius AP and the normal to S at P, then $dS \cos \theta$ is the projection of dS on a plane at right angles to OP. Using a theorem on similar areas

$$\frac{dS \cos \theta}{d\omega} = \frac{AP^2}{1^2}.$$

Hence

$$d\omega = \frac{dS \cos \theta}{r^2} \qquad (28)$$

$$\omega = \int \frac{\cos \theta}{r^2} dS \qquad (29)$$

where the integral is taken over the surface S and $AP = r$.

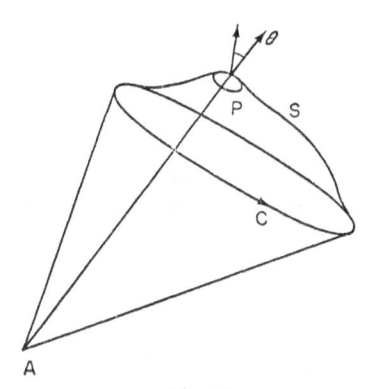

Fig. 13

When A and S are situated as in the figure, the angle θ is acute for all positions of P on S, and $d\omega$ is positive. If A is taken on the opposite side, the angle between dS and $\overrightarrow{AP}$ is obtuse, and $d\omega$ is negative.

Example. *Show that the solid angle at the vertex of a right circular cone of vertical angle 2θ is $2\pi(1 - \cos\theta)$.* Draw a sphere of unit radius with centre at vertex of cone. The area of the spherical cap within the cone is equal to the area cut off from the circumscribing cylinder, and is $2\pi(1 - \cos\theta)$.

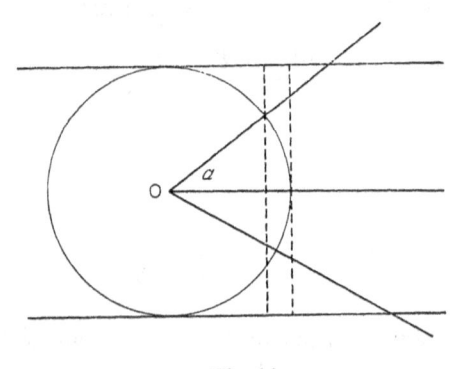

Fig. 14

Exercise. Show that the solid angle subtended by a closed surface at an external point is zero, at an internal point is 4π, and at a point on the surface is 2π.

To derive Gauss's theorem, we calculate the flux of the field across a closed surface S due to a mass m at A.

$$\mathbf{F} = -\gamma m \frac{\mathbf{r}}{r^3}$$

$$\mathbf{F} \cdot d\mathbf{S} = -\gamma m \frac{\mathbf{r} \cdot d\mathbf{S}}{r^3} = -\frac{\gamma m \, dS \cos\theta}{r^2} = -\gamma m \, d\omega.$$

SOME APPLICATIONS 129

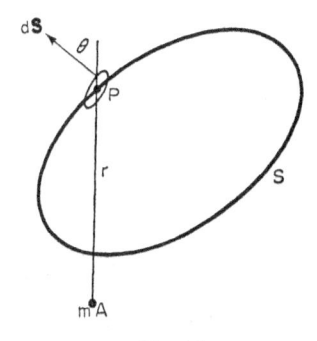

Fig. 15

Now

$$\int_S d\omega = 0 \quad \text{if } A \text{ is outside } S \tag{30}$$

$$= 4\pi \quad \text{if } A \text{ is inside } S \tag{31}$$

$$\therefore \quad N = \int_S \mathbf{F} \cdot d\mathbf{S} = 0 \quad \text{if } A \text{ is outside } S$$

$$= -4\pi\gamma m \quad \text{if } A \text{ is inside } S.$$

If there is a distribution of matter both inside and outside S, the masses outside give no contribution to the flux while each interior mass contributes $-4\pi\gamma$ times its mass. Hence for the whole distribution the flux is $-4\pi\gamma$ times the mass enclosed within S.

When the distribution is a volume distribution of density ρ, which may be variable, from Gauss's theorem may be deduced a differential equation for the field $\mathbf{F}$. Since

$$\int_S \mathbf{F} \cdot d\mathbf{S} = -4\pi\gamma \int \rho \, d\Omega,$$

transformation of the left-hand side into a volume integral by Gauss's transformation gives

$$\int_\Omega \text{div } \mathbf{F} \, d\Omega.$$

CONCISE VECTOR ANALYSIS

Hence
$$\int_\Omega (\text{div } \mathbf{F} + 4\pi\gamma\rho) \, d\Omega = 0.$$

This being true for every closed volume Ω, it may be inferred that at every point

$$\text{div } \mathbf{F} + 4\pi\gamma\rho = 0. \tag{32}$$

5.7. Gravitational potential

The field at P due to a particle m at O is

$$\mathbf{F} = -\gamma m \frac{\mathbf{r}}{r^3}.$$

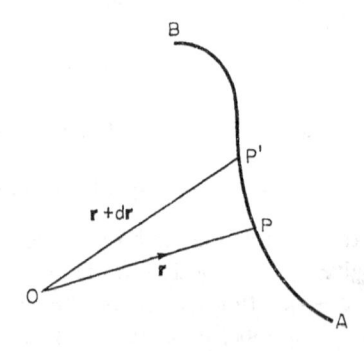

Fig. 16

If a unit particle is placed at P the force on it is $\mathbf{F}$. If this unit particle is displaced from P to a neighbouring point P' along the curve,

$$\text{work done} = \mathbf{F}.d\mathbf{r} = -\gamma m \frac{r \, dr}{r^3} = \gamma m \, d\left(\frac{1}{r}\right).$$

Hence work done in moving particle from A to B is

$$\gamma m \left(\frac{1}{r_A} - \frac{1}{r_B}\right), \tag{33}$$

SOME APPLICATIONS 131

which is independent of the path along which the particle is moved, and depends only on the initial and final points. Such a field is a potential field, and there exists a scalar function V such that

$$\mathbf{F} = \operatorname{grad} V, \tag{34}$$

where

$$V = \frac{\gamma m}{r} \tag{35}$$

gives the potential at P due to the mass m at O, and may also be defined as the work done in moving a unit particle from P to infinity.

For a system of n point masses m_i at r_i, the potential at P is

$$V = \sum_{i=1}^{n} \frac{\gamma m_i}{|\mathbf{r} - \mathbf{r}_i|}. \tag{36}$$

For a volume distribution of matter, the summation is replaced by an integral. With reference to the figure, the matter is divided into small elements $\rho(\mathbf{r}') \, d\Omega'$, and

$$V(r) = \int \frac{\gamma \rho(\mathbf{r}') \, d\Omega'}{|\mathbf{r}' - \mathbf{r}|}. \tag{37}$$

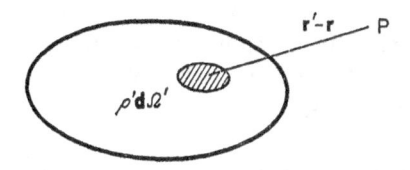

Fig. 17

For a surface distribution, and a line distribution

$$V(\mathbf{r}) = \int \frac{\gamma \sigma(\mathbf{r}') \, dS'}{|\mathbf{r} - \mathbf{r}'|} \quad \text{and} \quad V(\mathbf{r}) = \int \frac{\gamma \lambda(\mathbf{r}') \, dS'}{|\mathbf{r} - \mathbf{r}'|} \tag{38}$$

respectively.

When the point P is external to the distribution, there is no difficulty of the integrand becoming infinite, but when P is an internal point, the integrand becomes infinite and the question of the convergence of the integral needs to be considered.

At a point mass, V and $\mathbf{F}$ are both infinite. On a line distribution also V and $\mathbf{F}$ are generally singular functions. Within a volume distribution of density ρ, the integrals concerned converge, and V and $\mathbf{F}$ remain finite and continuous even within the distribution. Equation (32) requires V to satisfy a differential equation, namely

$$\text{div}(\text{grad } V) + 4\pi\gamma\rho = 0,$$

that is,

$$\nabla^2 V + 4\pi\gamma\rho = 0, \tag{39}$$

which is *Poisson's equation*.

For a surface distribution of finite density σ, the matter is in practice distributed to a small thickness on the surface, and the volume density of such a distribution is very large. Therefore $\nabla^2 V$ is very large, and though V remains finite and continuous at points on the surface, $\mathbf{F} = \text{grad } V$ is discontinuous on the surface. If F_t denotes the tangential component of $\mathbf{F}$ and F_n the component along the outward normal, F_t is found to be continuous on the two sides of the surface, but F_n suffers a discontinuity. The amount of discontinuity may be calculated by applying Gauss's theorem to a small cylinder of cross-section dS, infinitesimal height and axis along $\mathbf{n}$.

If F_{on} is normal component on the outer side of S and F_{in} on the inner Gauss's theorem gives

$$F_{on} \, dS - F_{in} \, dS = -4\pi\gamma\sigma \, dS,$$

whence

$$F_{on} - F_{in} = -4\pi\gamma\sigma. \tag{40}$$

In terms of the potential function, if V_o denotes potential in the outer region and V_i in the inner region

$$\frac{\partial V_o}{\partial n} - \frac{\partial V_i}{\partial n} = -4\pi\gamma\sigma, \tag{41}$$

where $\partial/\partial n$ denotes differentiation along the outward normal.

Given the potential as a function of position, the distribution producing it may be obtained from the above relations.

If V and grad V are continuous, then there will be no surface distribution. If grad V is discontinuous, the distribution would include a surface distribution the density σ of which may be obtained from the eqn. (41).

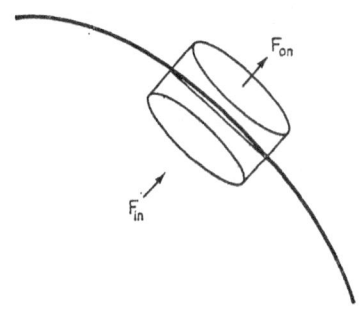

Fig. 18

On either side of the surface the potential function takes different forms, but these forms take equal values on the surface itself. The volume density at any point is obtained from V at any point by the relation

$$\rho = \frac{1}{4\pi}\nabla^2 V.$$

If the potential becomes infinite at a point, the mass there may be found by applying Gauss's theorem to a small volume surrounding the point. If the potential is infinite at all points on a line, the density λ may be obtained from Gauss's theorem by applying to a small cylinder with its axis along the line, and considering the limit as the cross-section of the cylinder is made to vanish.

Example. The potential outside and inside the sphere $x^2 + y^2 + z^2 - 2cz = 0$ being denoted by V_o and V_i and given that
$$V_o = 0$$
$$V_i = \pi\mu\gamma(x^2 + y^2 + z^2 - 2cz),$$

find the distribution of matter.

Since potential is continuous, the boundary is the surface on which $V = 0$, that is, the boundary is the sphere of centre $(0, 0, c)$ and radius c. Unit vector along normal is

$$\mathbf{n} = \left(\frac{x}{c}, \frac{y}{c}, \frac{z-c}{c}\right).$$

The volume density outside the sphere is
$$\rho_o = -\tfrac{1}{4}\pi\gamma\nabla^2 V_o = 0$$
The volume density inside the sphere is
$$\rho_i = -\tfrac{1}{4}\pi\gamma\nabla^2 V_i = -\tfrac{3}{2}\mu.$$
On the surface,
$$\sigma = \frac{1}{4\pi\gamma}\frac{\partial V_i}{\partial n}$$
$$= \frac{1}{4\pi\gamma}\left(\frac{\partial x}{\partial n}\frac{\partial V_i}{\partial x} + \frac{\partial y}{\partial n}\frac{\partial V_i}{\partial y} + \frac{\partial V_i}{\partial z}\right)$$
$$= \frac{1}{4\pi\gamma}\left(\frac{x}{c}\frac{\partial V}{\partial x} + \frac{y}{c}\frac{\partial V}{\partial y} + \frac{z-c}{c}\frac{\partial V}{\partial z}\right)$$
$$= \frac{\mu}{4c}\{2x^2 + 2y^2 + 2(z-c)^2\}$$
$$= \frac{\mu}{2}c.$$

Hence the distribution consists of a uniform sphere of volume density $-3\mu/2$, with a uniform surface coating of density $\tfrac{1}{2}\mu c$.

Given the distribution of matter, the determination of the potential function requires use of special methods depending on the complexity of the distribution. Sometimes the direct evaluation of the integral in (37) or (38) may be feasible. In some cases, first the gravitational field **F** may be obtained by applying Gauss's theorem, and thereafter the potential obtained by using

$$V(P) = \int_{\infty}^{P} \mathbf{F} \cdot d\mathbf{r}.$$

In still others, the potential may be obtained by solving Poisson's and Laplace's equations subject to appropriate boundary conditions. This is specially so when considerations of symmetry may reduce the number of variables that occur in the differential equations.

Example 1. *Show that the potential of a uniform circular disc, of radius a and surface density σ, at a point P on its axis at a distance z from the disc is*

$$2\pi\gamma\sigma\{\sqrt{(z^2 + a^2)} - z\}.$$

Find the potential at the vertex of a uniform solid right circular cone of density ρ, semi-vertical angle α and height h.

Divide the disc into circular rings. The circular ring of radius r and thickness dr has mass $2\pi r dr \sigma$, and its potential at P is

$$\gamma 2\pi r dr \sigma / \sqrt{(r^2 + z^2)}.$$

Hence, due to the whole disc,

$$V = 2\pi\gamma\sigma \int_0^a \frac{r \, dr}{\sqrt{(r^2 + z^2)}} = 2\pi\gamma\sigma[\sqrt{(a^2 + z^2)} - z].$$

The potential at the vertex of the cone may be obtained by dividing the cone into parallel circular discs. The disc at distance z from the vertex will have radius $z \tan \alpha$, thickness dz, and surface density $\rho \, dz$. The potential of the disc at O is

$$2\pi\gamma\rho \, dz[z \sec \alpha - z].$$

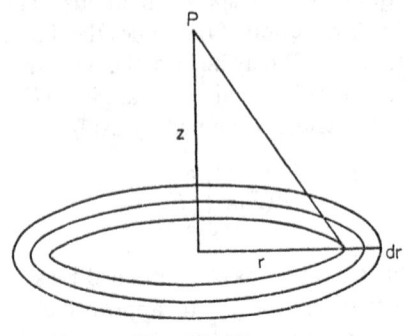

Fig. 19

Hence the potential at O of the cone is

$$V = \int_0^h 2\pi\gamma\rho(\sec \alpha - 1)z \, dz$$
$$= \pi\gamma\rho h^2(\sec \alpha - 1).$$

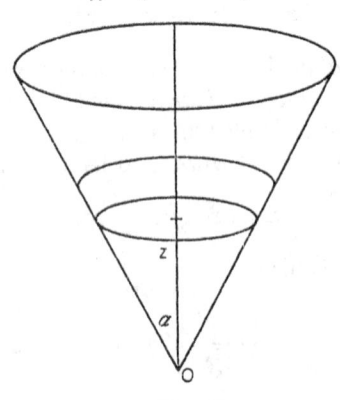

Fig. 20

Example 2. *Using Gauss's theorem, find the attraction and potential at all points due to a uniform solid sphere of density ρ and radius a.* There is spherical symmetry, and the field at P is radial

SOME APPLICATIONS 137

and is a function of r only. Suppose P is an internal point. Applying Gauss's theorem to a sphere of centre O and radius a,

$$F4\pi r^2 = -4\pi\gamma \cdot \tfrac{4}{3}\pi r^3 \rho$$

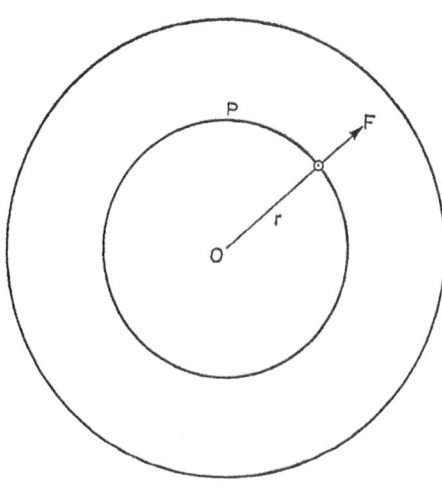

Fig. 21

Hence

$$F = -\tfrac{4}{3}\pi\gamma\rho r.$$

If P is an external point, Gauss's theorem gives

$$F \cdot 4\pi r^2 = -4\pi\gamma \cdot \tfrac{4}{3}\pi a^3 \rho,$$

that is

$$F = -\tfrac{4}{3}\pi\gamma a^3 \rho / r^2.$$

If V_o is potential at an external point, and V_i at internal point,

$$\frac{dV_o}{dr} = F = -\frac{\tfrac{4}{3}\pi\gamma a^3 \rho}{r^2}$$

Integrating,

$$V_o = \tfrac{4}{3}\pi\gamma a^3 \rho \frac{1}{r} + C.$$

Since $V_o \to 0$ as $r \to \infty$, $C = 0$.

$$\frac{dV_i}{dr} = -\tfrac{4}{3}\pi\gamma\rho r.$$

Integrating

$$V_i = -\tfrac{2}{3}\pi\gamma\rho r^2 + D.$$

On $r = a$, $V_o = V_i$.

$$\tfrac{4}{3}\pi\gamma a^2\rho = -\tfrac{2}{3}\pi\gamma\rho a^2 + D, \qquad D = 2\pi\gamma a^2\rho.$$

Hence

$$V = \tfrac{4}{3}\pi\gamma\rho a^3/r \qquad r > a$$
$$= \tfrac{2}{3}\pi\gamma\rho(3a^2 - r^2) \qquad r < a.$$

Example 3. If

$$V = f(r)\cos\theta$$

is a solution of Laplace's equation in spherical polar coordinates r, θ, ϕ, *show that*

$$f(r) = Ar + \frac{B}{r^2}$$

where A, B are constants.

A surface distribution on the sphere $r = a$ *has density* $k\cos\theta$, *where k is a constant. Given that the potentials inside and outside the sphere are of the form* $f(r)\cos\theta$, *determine the potential at all points.*

In spherical polar coordinates Laplace's equation is

$$\frac{1}{r^2}\left[\frac{\partial}{\partial r}\left(r^2\frac{\partial V}{\partial r}\right) + \frac{1}{\sin\theta}\frac{\partial}{\partial\theta}\left(\sin\theta\frac{\partial V}{\partial\theta}\right) + \frac{1}{\sin^2\theta}\frac{\partial^2 V}{\partial\phi^2}\right] = 0.$$

If $V = f(r)\cos\theta$ is a solution,

$$\frac{\partial}{\partial r}\left(r^2\frac{\partial f}{\partial r}\cos\theta\right) + \frac{1}{\sin\theta}\frac{\partial}{\partial\theta}\left[-\sin^2\theta f(r)\right] = 0$$

Dividing by $\cos\theta$,

$$\frac{d}{dr}\left(r^2\frac{df}{dr}\right) - 2f = 0.$$

If we change from variable r to t where $r = e^t$, the equation becomes

$$\frac{d^2 f}{dt^2} + \frac{df}{dt} - 2f = 0.$$

This is a linear equation with constant coefficients, and the auxiliary equation is

$$\lambda^2 + \lambda - 2 = 0$$

which has roots $1, -2$. Hence

$$f = Ae^t + Be^{-2t} = Ar + \frac{B}{r^2}.$$

V_o, V_i have to be determined to satisfy the following conditions:

(1) V_o, V_i have the form $f(r) \cos \theta$.

(2) $\nabla^2 V_o = 0$ for all $r > a$, $\nabla^2 V_i = 0$ for all $r < a$.

(3) $V_o \to 0$ as $r \to \infty$.

(4) V_o, V_i are finite everywhere.

(5) $V_o = V_i$ on $r = a$.

(6) $\dfrac{\partial V_o}{\partial r} - \dfrac{\partial V_i}{\partial r} = -4\pi\gamma\sigma$ on $r = a$.

From (1),

$$V_o = \left(Ar + \frac{B}{r^2}\right)\cos \theta \qquad r > a,$$

$$V_i = \left(Cr + \frac{D}{r^2}\right)\cos \theta \qquad r < a.$$

From (3),

$$A = 0.$$

From (4),

$D = 0$, since the potential should be finite at the origin.

From (5),
$$\frac{B}{a^2} \cos \theta = Ca \cos \theta, \text{ that is, } B = Ca^3,$$

From (6)
$$-\frac{2B}{a^3} \cos \theta - C \cos \theta = -4\pi\gamma\sigma = -4\pi\gamma k \cos \theta$$

whence
$$2B + Ca^3 = 4\pi\gamma k a^3.$$

Thus
$$B = Ca^3 = \tfrac{4}{3}\pi\gamma k a^3.$$

Hence
$$V_o = \tfrac{4}{3}\pi\gamma k a^3 \frac{\cos \theta}{r^2} \qquad r > a,$$

$$V_i = \tfrac{4}{3}\pi\gamma k r \cos \theta \qquad r < a.$$

5.8. Equipotential surfaces

The surfaces $V(x, y, z) = \textit{constant}$ are a family of surfaces, on each of which the potential is constant. At any point P, there is just one equipotential surface which passes through it. The field vector **F** at P is normal to this equipotential surface, and in general lines of force cut equipotential surfaces orthogonally.

It is of interest to investigate the condition that a given family of surfaces

$$f(x, y, z) = \text{constant} \tag{42}$$

may be a possible set of equipotentials in free space. When this condition is satisfied, V is constant whenever f is constant, and hence there is a functional relation between them which we may write as

$$V = \psi\{f(x, y, z)\}. \tag{43}$$

SOME APPLICATIONS

Then
$$\frac{\partial V}{\partial x} = \psi'(f)\frac{\partial f}{\partial x}, \qquad \frac{\partial^2 V}{\partial x^2} = \psi''(f)\left(\frac{\partial f}{\partial x}\right)^2 + \psi'(f)\frac{\partial^2 f}{\partial x^2}$$

$$\nabla^2 V = \psi''(f)\left\{\left(\frac{\partial f}{\partial x}\right)^2 + \left(\frac{\partial f}{\partial y}\right)^2 + \left(\frac{\partial f}{\partial z}\right)^2\right\} + \psi'(f)\nabla^2 f$$

Since $\nabla^2 V = 0$ in free space

$$\frac{\nabla^2 f}{(\partial f/\partial x)^2 + (\partial f/\partial y)^2 + (\partial f/\partial z)^2} = -\frac{\psi''(f)}{\psi'(f)}$$
$$= \text{a function of } f, \text{ say } G(f). \qquad (44)$$

Hence the required condition is that

$$\nabla^2 f/|\operatorname{grad} f|^2$$

should be expressible as a function of $f(x, y, z)$ only. When this condition is satisfied, the potential function may be obtained by integrating

$$\frac{\psi''(f)}{\psi'(f)} = -G(f).$$

Thus
$$\log \psi'(f) = C - \int G(f)\, df,$$

or
$$\psi'(f) = B \exp\left\{-\int G(f)\, df\right\}$$

Hence
$$V = \psi(f) = B \int \exp\left\{-\int G(f)\, df\right\} df + C. \qquad (45)$$

Example. Show that the system of coaxial cylinders

$$x^2 + y^2 + 2\lambda x + c^2 = 0,$$

where c is constant and λ variable, can form a system of equipotential surfaces in free space, and determine the potential function.

Write the equation of the surfaces in the form

$$f(x, y, z) = \frac{x^2 + y^2 + c^2}{x} = \text{constant}.$$

$$\frac{\partial f}{\partial x} = 1 - \frac{y^2 + c^2}{x^2}, \qquad \frac{\partial^2 f}{\partial x^2} = \frac{2(y^2 + c^2)}{x^3}$$

$$\frac{\partial f}{\partial y} = \frac{2y}{x}, \qquad \frac{\partial^2 f}{\partial y^2} = \frac{2}{x}.$$

$$\frac{\nabla^2 f}{|\operatorname{grad} f|^2} = \frac{\dfrac{2(y^2 + c^2)}{x^3} + \dfrac{2}{x}}{\left(1 - \dfrac{y^2 + c^2}{x^2}\right)^2 + \dfrac{4y^2}{x^2}}$$

$$= \frac{2f}{f^2 - 4c^2}.$$

Hence the condition is satisfied. Further

$$\frac{\psi''(f)}{\psi'(f)} = -\frac{2f}{f^2 - 4c^2}. \quad \log \psi'(f) = A - \log(f^2 - 4c^2).$$

$$\psi'(f) = \frac{B}{f^2 - 4c^2} = \frac{B}{4c}\left[\frac{1}{f - 2c} - \frac{1}{f + 2c}\right].$$

$$V = \psi(f) = C \log \frac{f - 2c}{f + 2c} + D$$

$$= C \log \frac{(x - c)^2 + y^2}{(x + c)^2 + y^2} + D.$$

5.9. Green's theorems

If ϕ, ψ are two single-valued continuous functions with continuous first and second order derivatives, and Ω the volume bounded by a closed surface S,

$$\int_\Omega (\phi \nabla^2 \psi - \psi \nabla^2 \phi) \, d\Omega = \int_S \left(\phi \frac{\partial \psi}{\partial n} - \psi \frac{\partial \phi}{\partial n}\right) dS, \qquad (46)$$

$$\int_\Omega \{\phi\nabla^2\phi + (\text{grad } \phi)^2\} \, d\Omega = \int_S \phi \frac{\partial \phi}{\partial n} \, dS. \tag{47}$$

In the relation

$$\text{div}(\phi \mathbf{A}) = \phi \text{ div } \mathbf{A} + (\text{grad } \phi).\mathbf{A}$$

take $\mathbf{A} = \text{grad } \psi$

$$\text{div}(\phi \text{ grad } \psi) = \phi\nabla^2\psi + (\text{grad } \phi).(\text{grad } \psi). \tag{48}$$

Interchanging ϕ and ψ

$$\text{div}(\psi \text{ grad } \phi) = \psi\nabla^2\phi + (\text{grad } \psi).(\text{grad } \phi). \tag{49}$$

Subtracting (49) from (48),

$$\text{div}(\phi \text{ grad } \psi - \psi \text{ grad } \phi) = \phi\nabla^2\psi - \psi\nabla^2\phi.$$

Integrating over the volume Ω, the right-hand side gives

$$\int_\Omega (\phi\nabla^2\psi - \psi\nabla^2\phi) \, d\Omega,$$

and the left-hand side gives

$$\int_\Omega \text{div}(\phi \text{ grad } \psi - \psi \text{ grad } \phi) \, d\Omega,$$

which can be transformed by Gauss's transformation into the surface integral

$$\int_S (\phi \text{ grad } \psi - \psi \text{ grad } \phi)_n \, dS$$

which is

$$\int_S \left(\phi \frac{\partial \psi}{\partial n} - \psi \frac{\partial \phi}{\partial n}\right) dS.$$

Hence we obtain the eqn. (46), which we call the first form of Green's theorem.

In eqn. (48) putting $\psi = \phi$ and integrating over the volume Ω, the right-hand side gives

$$\int_\Omega \{\phi \nabla^2 \phi + (\text{grad } \phi)^2\} \, d\Omega,$$

and the left-hand side gives after transformation into a surface integral by Gauss's transformation

$$\int_S \phi \frac{\partial \phi}{\partial n} \, dS.$$

Hence we obtain the eqn. (47), which is the second form of Green's theorem.

These two forms of Green's theorem are very useful in potential theory and in particular in the proof of many general results about potential functions. The second form, for example, is used for proving the existence of a unique potential function when it is required to satisfy certain conditions.

Gauss's theorem for a volume distribution may be derived from the first form of Green's theorem by taking $\phi = 1$ and ψ as the potential function V. Then $\nabla^2 \phi = 0$, $\nabla^2 \psi = \nabla^2 V = -4\pi\gamma\rho$, and equation (46) gives

$$\int_\Omega (-4\pi\gamma\rho) \, d\Omega = \int_S \frac{\partial \psi}{\partial n} \, dS.$$

that is

$$-4\pi\gamma \int_\Omega \rho \, d\Omega = \int_S F_n \, dS.$$

Another theorem that may be derived from the first form of Green's theorem is called *Green's Reciprocal Theorem*, which we may state in the following form for finite volume distributions.

If a volume distribution of density ρ gives rise to a potential function V, and a second distribution of density ρ' has potential V',

$$\int_\Omega \rho V' \, d\Omega = \int_\Omega \rho' V \, d\Omega. \tag{50}$$

Take $\phi = V$ and $\psi = V'$. Then $\nabla^2 \phi = -4\pi\gamma\rho$, $\nabla^2 \psi = -4\pi\gamma\rho'$. Apply first form of Green's theorem to the whole of space bounded by the infinite sphere Σ of radius R where R tends to infinity. The volume integral then is

$$\int_\Omega \{V(-4\pi\gamma\rho') - V'(-4\pi\gamma\rho)\} \, d\Omega$$
$$= -4\pi\gamma\left[\int \rho'V \, d\Omega - \int \rho V' \, d\Omega\right].$$

The surface integral vanishes, since on the infinite sphere, V, V' are each of order $1/R$, $\partial V/\partial n$, $\partial V'/\partial n$ are of order $1/R^2$, and dS is over $R^2 \, d\omega$ where ω is a solid angle. Hence the surface integral is of the order $1/R$, and vanishes in the limit when R tends to infinity. Hence from Green's theorem we have the eqn. (50).

For two systems of point masses, by analogy with (50), it is found that

$$\sum mV' = \sum m'V \tag{51}$$

but the proof would require an extension of Green's theorem to regions of space where V and V' are not continuous.

Exercise V

1. Forces X, Y, Z act along the three lines given by the equations

$$y = 0, z = c; \quad z = 0, x = a; \quad x = 0, y = b;$$

prove that the pitch of the equivalent wrench is

$$(aYZ + bZX + cXY)/(X^2 + Y^2 + Z^2).$$

If the wrench reduces to a single force, show that the line of action of the force must lie on the hyperboloid

$$(x - a)(y - b)(z - c) - xyz = 0$$

2. Three edges OA, OB, OC of a tetrahedron are equal in length, mutually perpendicular and right-handedly related in that order. D is the mid-point of AB, and OD is of length p. Show that the set of forces equivalent to forces P and Q localized in $\overrightarrow{AB}$ and $\overrightarrow{OC}$ respectively, together with couples λP and λQ, is equivalent to a single force if

$$\lambda = -pPQ/(P^2 + Q^2).$$

Show also that the force cuts OD at right angles at the point which divides OD in the ratio $P^2:Q^2$. [London S. 1957]

3. A force P acts along a generator

$$\frac{x - a\cos\theta}{a\sin\theta} = \frac{y - b\sin\theta}{-b\cos\theta} = \frac{z}{c}$$

of the hyperboloid $x^2/a^2 + y^2/b^2 - z^2/c^2 = 1$. Find its force and couple components with respect to the axes of the hyperboloid.

If the force P and a given force **Q** whose force components are X, Y, Z reduce to a single force for all values of θ, prove that

$$X^2/a^2 + Y^2/b^2 - Z^2/c^2 = 0 \qquad \text{[Ceylon S. 1943]}$$

4. Show that there are three points on the curve

$$\mathbf{r} = \mathbf{a}u^3 + \mathbf{b}u^2 + \mathbf{c}u + \mathbf{d}$$

the osculating planes at which pass through the origin, and that they lie in the plane

$$[\mathbf{r}, \mathbf{b}, \mathbf{c}] = 3[\mathbf{r}, \mathbf{a}, \mathbf{d}] \qquad \text{[cf. Oxford II 1957]}$$

5. Let **r** be the position vector of a variable point P of a twisted space curve Γ and let **n** be the principal normal to Γ and K the curvature of Γ at P. If Γ_1 is the locus of points P_1 with position vectors

$$\mathbf{r}_1 = \mathbf{r} + K^{-1}\mathbf{n}$$

show that the tangents to Γ and to Γ_1, at corresponding points P and P_1, are perpendicular to each other.

If K is constant show that the curvature K_1 of Γ_1 and the product $\tau\tau_1$ of the torsions of Γ and of Γ_1 at corresponding points are constant.

Find K, τ, τ_1 when Γ is given in terms of the parameter θ by

$$r = (\cos\theta, \sin\theta, \theta\cot\beta) \qquad \text{[London S. 1960]}$$

6. A point O moves along a twisted curve with velocity **v**, and the coordinates of P, referred to the principal axes of the curve at O, are ξ, η, ζ. Show that the velocities of P parallel to the moving axes are

$$v + \dot\xi - \frac{\eta v}{\rho}, \quad \dot\eta - \frac{\zeta v}{\sigma} + \frac{\xi v}{\rho}, \quad \dot\zeta + \frac{\eta v}{\sigma},$$

where ρ and σ are the radii of curvature and torsion.

Prove that, if ρ lies at a constant distance c from O along the principal normal, OP will be the bi-normal to the locus of P if

$$\sigma^2\rho = c(\rho^2 + \sigma^2). \qquad \text{[Camb. M.T. (1919)]}$$

7. The vector **F** indicates one member of a rigid framework. The framework is now rotated about an axis along the unit vector **i** through an angle α. Show that **F** is turned into **G** given by

$$\mathbf{G} - \mathbf{F} = \{\boldsymbol{\omega} \times \mathbf{F} + \tfrac{1}{2}\boldsymbol{\omega} \times (\mathbf{F} \times \boldsymbol{\omega})\}/(1 + \tfrac{1}{4}\omega^2)$$

where

$$\boldsymbol{\omega} = 2\mathbf{i}\tan\tfrac{1}{2}\alpha. \qquad \text{[Camb. P.N.S. 1956]}$$

SOME APPLICATIONS   147

8. The position vector **x** of a point P satisfies the differential equation
$$\frac{d\mathbf{x}}{dt} = \boldsymbol{\omega} \times \mathbf{x},$$
where $\boldsymbol{\omega}$ is a fixed vector. Show that P lies on a fixed sphere and also in a fixed plane.

Deduce that P moves in a circle and show that it describes the circle with constant speed. Find the angular velocity with which the circle is described.

[Camb. P.N.S. 1956]

9. The motion of a lamina in its own plane is specified at a certain instant by the velocity **V** of a point P of the lamina, and its angular velocity $\boldsymbol{\omega}$ about P. Prove that the position vector of the instantaneous centre I relative to P is $(\boldsymbol{\omega} \times \mathbf{V})/\omega^2$.

A circular disc, moving in its own plane, spins with angular velocity $\boldsymbol{\omega}$, while its centre C describes a circle of centre O and radius R with angular velocity $\boldsymbol{\Omega}$. Show that when the position vector of C relative to O is **R**, that of the instantaneous centre is $(1 - \Omega/\omega)\mathbf{R}$. Hence find the space and body centrodes (loci of I in space and on the disc) when Ω/ω is constant.

[Camb. M.T. 1954]

10. A rigid body S has spin $\boldsymbol{\omega}$ and a particle A of S has velocity **v**. Show that every particle P of S with velocity vector parallel to $\boldsymbol{\omega}$ lies on the line
$$\overrightarrow{AP} = \frac{\boldsymbol{\omega} \times \mathbf{v}}{\omega^2} + \mu\boldsymbol{\omega},$$
μ being an arbitrary scalar.

The instantaneous velocities of particles at the points (a, o, o), $(o, a/\sqrt{3}, 0)$ $(0, 0, 2a)$ of a rigid body are $(u, 0, 0)$, $(u, 0, v)$, $(u + v, -v\sqrt{3}, v/2)$ respectively, referred to a rectangular cartesian frame. Find the magnitude and direction of the spin of the body and the point at which the central axis cuts the xz-plane.

[London S. 1959]

11. Two right circular cones have the same vertex O and the same axis OX, and have semivertical angles α, $\beta (\beta > \alpha)$. They rotate about OX with angular velocities ω and Ω, respectively, and a sphere of centre C rolls without slipping in the space between the cones. Show that the plane COX rotates with angular velocity

$$(\omega \sin \alpha + \Omega \sin \beta)/(\sin \alpha + \sin \beta).$$

12. A rigid body has simultaneous angular velocities about the lines
$$y = a, z = -a; \quad z = a, x = -a; \ x = a, y = -a.$$
If the resultant screw motion has a pitch a, show that its axis lies in the plane
$$x + y + z = 0.$$

13. Show that potential at P of a uniform rod AB of length $2a$ and mass λ per unit length is
$$\gamma\lambda \log \frac{r + r' + 2a}{r' + r' - 2a}$$
where $r = AP$, $r' = BP$.

Show that the equipotential surfaces are a family of spheroids.

14. The density of an infinite circular cylinder of radius a at a distance r from the axis is $k\sqrt{(a^2 - r^2)}$, and M is the mass per unit length of the cylinder. Show that the attraction at an internal point is

$$\frac{2\gamma M}{a^3 r}[a^3 - (a^2 - r^2)^{3/2}],$$

and find the attraction at external points.

15. If the gravitational potential is
$$V = Cx, \qquad r < a$$
$$= \frac{Ca^3 x}{(x^2 + y^2 + z^2)^{3/2}} \qquad r > a,$$

where $r = \sqrt{(x^2 + y^2 + x^2)}$ and C a constant, find the distribution.

16. If $V = \tfrac{4}{3}\pi\gamma\rho\left[\dfrac{2a}{r} + \dfrac{a^5}{5r^5}(2z^2 - x^2 - y^2)\right]$, $\qquad r > a$

$= \tfrac{4}{3}\pi\gamma\rho[\tfrac{3}{2}a^2 - \tfrac{1}{2}r^2 + \tfrac{1}{5}(2z^2 - x^2 - y^2)]$, $\qquad r > a$

where $r = \sqrt{(x^2 + y^2 + z^2)}$, find the distribution of matter.

17. The axes being rectangular, determine what volume and surface distributions of matter will give rise to potential

$$Kxyz(a - x - y - z)$$

where K and a are constants, at all points within the tetrahedron bounded by the coordinate planes and the planes $x + y + z = a$, and potential zero at all points outside the tetrahedron.

Show that the total mass of the matter contained within the tetrahedron is $Ka^5/80\pi\gamma$. [London S. 1938]

18. The equation $f(r, \theta) = \lambda$ where r, θ, z are cylindrical coordinates and λ is a variable, represents a family of equipotential surfaces in free spaces.

Prove that

$$\left[\frac{\partial^2 f}{\partial r^2} + \frac{1}{r^2}\frac{\partial^2 f}{\partial \theta^2} + \frac{1}{r}\frac{\partial f}{\partial r}\right] \bigg/ \left[\left(\frac{\partial f}{\partial r}\right)^2 + \frac{1}{r^2}\left(\frac{\partial f}{\partial \theta}\right)^2\right]$$

is a function of λ only.

Show that the system of surfaces

$$r^8 + a^8 - 2a^4 r^4 \cos 4\theta = \lambda$$

where r, θ, z are cylindrical coordinates, can form a system of equipotential surfaces and that the corresponding potential function is

$$V = A \log \lambda + B$$

where A and B are arbitrary constants. [London S. 1946]

19. Show that the family of cylinders

$$(x^2 + y^2)^3 - 2a^3(x^3 - 3xy^2) + a^6 = \text{const.}$$

is a possible form of equipotential surfaces, and find the corresponding potential. [London S. 1938]

20. Show that the family of spheroids

$$\frac{x^2 + y^2}{a^2 + \lambda} + \frac{z^2}{c^2 + \lambda} = 1$$

where λ is a variable parameter, is a possible form of equipotential surfaces, and determine the potential function.

21. If

$$V = r^n e^{im\phi} F(\theta)$$

satisfies Laplace's equation in spherical polar coordinates r, θ, ϕ where n and m are integers, and $F(\theta)$ is a function of θ only, show that $F(\theta)$ satisfies the differential equation

$$\frac{d}{d\mu}\left[(1 - \mu^2)\frac{dF}{d\mu}\right] + \left[n(n + 1) - \frac{m^2}{1 - \mu^2}\right] F = 0$$

where $\mu = \cos \theta$.

22. Two functions of position ϕ_1 and ϕ_2 are such that at all points inside a surface S, $\nabla^2 \phi_1 = 0$, $\nabla^2 \phi_2 = 0$. On the surface $\dfrac{\partial \phi_1}{\partial n} = \dfrac{\partial \phi_2}{\partial n}$ where $\dfrac{\partial}{\partial n}$ denotes differentiation along the outward normal. By applying the divergence theorem to $(\phi_1 - \phi_2)\,\text{grad}(\phi_1 - \phi_2)$, or otherwise, show that ϕ_1 and ϕ_2 differ by a constant. [Camb. N.S. 1949]

23. If vectors **E** and **H** are functions of position and time, and satisfy Maxwell's equations

$$\text{div } \mathbf{E} = 0, \qquad \text{div } \mathbf{H} = 0$$

$$\text{curl } \mathbf{E} = -\frac{1}{c}\frac{\partial H}{\partial t}, \qquad \text{curl } \mathbf{H} = \frac{1}{c}\frac{\partial E}{\partial t},$$

show that **E** satisfies

$$\nabla^2 \mathbf{E} = \frac{1}{c^2}\frac{\partial^2}{\partial t^2}\mathbf{E}.$$

Show also that Maxwell's equations are satisfied by

$$\mathbf{E} + i\mathbf{H} = i\,\text{curl}\,\frac{\partial \mathbf{S}}{\partial t} + \text{curl curl }\mathbf{S}$$

where **S** is a complex vector function of position and time satisfying

$$\nabla^2 \mathbf{S} = \frac{1}{c^2}\frac{\partial^2}{\partial t^2}\mathbf{S}. \qquad \text{[Lond. G.II. 1958]}$$

24. Assuming Green's theorem

$$\int_\Omega (\phi \nabla^2 \psi - \psi \nabla^2 \phi) \, d\Omega = \int_S \left(\phi \frac{\partial \psi}{\partial n} - \psi \frac{\partial \phi}{\partial n} \right) dS,$$

prove that the value of ψ_P, at a point inside S, of a function which satisfies the relation $\nabla^2 \psi = 0$ inside S, is given by the relation

$$\psi_P = \frac{1}{4\pi} \int_S \frac{1}{r} \frac{\partial \psi}{\partial n} \, dS - \frac{1}{4\pi} \int_\Omega \psi \frac{\partial}{\partial n} \left(\frac{1}{r} \right) dS,$$

where r is the distance of dS from P.

Verify the result by evaluating each of these integrals in the case when P is the origin, S is the sphere

$$x^2 + y^2 + z^2 = a^2$$

and ψ is the function

$$yz + zx + xy. \qquad \text{[Camb. P.N.S. 1949]}$$

INDEX

Acceleration vector 46
Addition of vectors 3
Angular velocity 92, 117, 119
Associative law 4

Binormal 114

Cartesian components 17, 26, 59
Centroid 11, 12
Circulation 70, 88, 89, 99
Commutative law 16, 119
Component 5, 19
Conservative field 69
Coplanar vectors 29
Couple 105
Cross product 22
Curl 88
Curvature 116
Curvilinear coordinates 96
Curvilinear integral 59
Cyclic 24, 29, 98
Cycloid 62
Cylindrical polar coordinates 98

Del ∇ 94
Density 125, 132
Derivative 38
Determinant 29
Differentiable scalar field 45
Differential vector operator 94
Differentiation of vectors 41
Direction 1
Direction cosines 8, 9
Directional derivative 48
Distributive law 6, 18, 24
Divergence 81

Divergence theorem 102
Dot product 16
Dynamics 113

Electric intensity 1
Electrostatic force field 45
Electrostatic potential 46
Equilibrium 47
Equipotential surface 46, 140
Equivalent 103

Field 44, 125
Fluid motion 45
Flux 77, 82, 101, 126
Force 1, 103
Frames of reference 7
Free space 141
Free vector 1
Functions of a vector 44

Gauss's theorem 124, 125, 129, 132, 137
Gauss's transformation 85, 129, 144
Geometry 39, 52
Gradient 50, 99
Gravitational force field 45, 46, 69
Gravitational potential 130
Green's reciprocal theorem 144
Green's theorem 100, 142, 144

Inner Product 16
Instantaneous axis 121
Invariance 110
Inverse square law 125

INDEX

Jacobian 80

Kinematics 122

Laplace's equation 96, 138
Laplacian operator ∇^2 96
Left-handed system 31
Leibnitz's theorem 15
Level surface 45, 51
Line integrals 58, 65, 93
Lines of force 48, 140
Localized vector 27

Magnitude of a vector 1, 3
Magnetic field 48
Magnetostatic field 48
Map of a field 45
Maxwell's equations 149
Moment 27, 105
Momentum 1, 103
Multiplication of vectors 6
Mutual moment 30

Nabla ∇ 94
Newton's law 63
Normal 52
Null vector 6

Operator 94
Orthogonal 140
Orthogonal curvilinear coordinates 96
Orthogonal triad 36, 115
Osculating plane 114, 115, 146
Outer product 22
Outward normal 74, 82

Parallelogram law 1, 104
Parameter 43, 96
Partial derivative 48
Path of integration 58, 66
Pitch 109

Point function 101
Point vector 44
Poinsot's central axis 107
Poisson's equation 96, 132
Polygon of forces 5
Position vector 8, 105
Potential 124, 133
Potential energy 46, 69
Potential function 141, 144
Potential vector 68, 70
Principal normal 114
Product 16–29
  scalar 16
  scalar triple 28
  vector 22
  vector triple 29
Projection 9

Radius of curvature 115
Radius of torsion 115
Reciprocal sets 36
Rectangular coordinate system 43, 44
Resolution of a vector 7
Resultant 5
Riemann integral 59
Right-handed coordinate systems 31
Rigid body 1
Rotatory 92

Scalar 1
Scalar field 44
Scalar function 44, 65
Scalar point function 58
Screw 121
Similar triangle 7
Sink 82, 124
Sliding vector 1
Solenoidal 84
Solid angle 127
Source 82
Space curve 113, 116
Spherical polar coordinates 98, 138
Spinors 1

INDEX

Stokes's theorem 92, 93, 100
Subtraction of vectors 6
Surface integrals 71, 93
Serret–Frenet formulae 115

Tangent vector 42, 124
Tensors 1
Tied vector 1
Three dimensional space 43
Transformation 87
Transmissibility 104
Triad 31
Triangle rule 3
Triple integral 78
Triple products 28, 31
Tubes of force 48

Unit vector 22
Unlocalized vector 1

Vector 1
  addition 3
  algebra 95
  field 44, 65, 68
  function 38
  line integral 61
Vector product 22
  calculus 38, 95
  moment 106
Velocity field 82, 92
Volume 78, 142
Volume integrals 78
Vorticity 92

Weighted mean centre 12
Work 131
Wrench 109